Principles of Ecology

Principles of Ecology

Curtis Carson

SYRAWOOD
PUBLISHING HOUSE

New York

Published by Syrawood Publishing House,
750 Third Avenue, 9th Floor,
New York, NY 10017, USA
www.syrawoodpublishinghouse.com

Principles of Ecology
Curtis Carson

International Standard Book Number: 978-1-68286-820-1 (Hardback)

Cataloging-in-Publication Data

Principles of ecology / Curtis Carson.
    p. cm.
Includes bibliographical references and index.
ISBN 978-1-68286-820-1
1. Ecology. 2. Environmental sciences. I. Carson, Curtis.
QH541 .P75 2019
577--dc23

TABLE OF CONTENTS

Preface VII

Chapter 1   **An Introduction to Ecology and Ecosystem** 1
- Ecology 1
- Abiotic Component 42
- Biotic Component 44
- Ecological Trap 46
- Ecosystem 46
- Ecosystem Management 78
- Ecological Resilience 78
- Ecosystem Services 85

Chapter 2   **Energy Flow in Ecosystems** 89
- Ecosystem Ecology 89
- Energy Flow 90
- Ecosystem Engineer 100
- Productivity 101
- Nutrient Cycle 102
- Primary Producers 112
- Food Web 113
- Food Chain 126
- Ecological Pyramid 127
- Ecological Efficiency 129
- Primary Production 132

Chapter 3   **Population Ecology** 134
- Population Growth 154
- Population Dynamics 160
- Abundance 165
- r/K Selection Theory 166

Chapter 4   **Understanding the Diverse Biogeochemical Cycles** 172
- Biogeochemical Cycle 172
- Water Cycle 175
- Carbon Cycle 183

- Oxygen Cycle 191
- Nitrogen Cycle 195
- Phosphorus Cycle 202
- Sulfur Cycle 205

Permissions

Index

PREFACE

Ecology is a branch of biology concerned with the study of interactions and interrelationships between organisms and their environment, as well as with other organisms. Ecosystems are vast systems of organisms, their communities, and the environmental factors that have an influence on these. Several processes control the flux of matter and energy through an environment, such as pedogenesis, nutrient cycling, primary production and niche construction. The study of ecology focuses on such processes, as well as ecological succession, distribution of organisms and biodiversity, among others. Ecosystems sustain life, regulate climate and produce economically crucial materials, such as biomass. The regulation of water filtration, erosion control, flood protection, global biogeochemical cycles, etc. is also sustained by the ecosystem. The book aims to shed light on some of the unexplored aspects of ecology. Some of the diverse topics covered in this book address the varied branches that fall under this category. It aims to serve as a resource guide for students and experts alike and contribute to the growth of the discipline.

A short introduction to every chapter is written below to provide an overview of the content of the book:

Chapter 1- An ecosystem is a community of biotic and abiotic components. The study of the interactions between organisms and their environment is under the domain of ecology. This is an introductory chapter, which will introduce briefly all the significant aspects of an ecosystem and its ecology, such as ecological trap, ecological resilience, ecosystem services and ecosystem management; **Chapter 2**- In ecology, the flow of energy through a food chain is referred to as the energy flow. The food chain consists of the primary consumers or herbivores, carnivores or secondary consumers, tertiary consumers and decomposers. This chapter discusses in detail the process of nutrient cycle and energy flow in ecosystems. Some of the significant topics encompassed herein include food web, food chain, ecological pyramid, ecological efficiency, etc.; **Chapter 3**- Population ecology is a sub-field of ecology. It studies the interactions between species populations and the environment. The concepts of population growth, population dynamics, abundance and r/K selection theory are fundamental to the development of population ecology. All such topics have been extensively discussed in this chapter; **Chapter 4**- A biogeochemical cycle refers to the cyclic pathway by which chemical substances move through the biotic and abiotic compartments of the Earth. Carbon, oxygen, sulfur, phosphorus, nitrogen and water cycles are significant processes of the Earth, which have been discussed in detail in this chapter.

I extend my sincere thanks to the publisher for considering me worthy of this task. Finally, I thank my family for being a source of support and help.

Curtis Carson

Chapter 1

An Introduction to Ecology and Ecosystem

An ecosystem is a community of biotic and abiotic components. The study of the interactions between organisms and their environment is under the domain of ecology. This is an introductory chapter, which will introduce briefly all the significant aspects of an ecosystem and its ecology, such as ecological trap, ecological resilience, ecosystem services and ecosystem management.

Ecology

Ecology is the scientific study of the distributions, abundance and relations of organisms and their interactions with the environment. Ecology includes the study of plant and animal populations, plant and animal communities and ecosystems. Ecosystems describe the web or network of relations among organisms at different scales of organization. Since ecology refers to any form of biodiversity, ecologists research everything from tiny bacteria's role in nutrient recycling to the effects of tropical rain forest on the Earth's atmosphere. The discipline of ecology emerged from the natural sciences in the late 19th century. Ecology is not synonymous with environment, environmentalism, or environmental science. Ecology is closely related to the disciplines of physiology, evolution, genetics and behavior.

Like many of the natural sciences, a conceptual understanding of ecology is found in the broader details of study, including:

- life processes explaining adaptations
- distribution and abundance of organisms
- the movement of materials and energy through living communities
- the successional development of ecosystems, and
- the abundance and distribution of biodiversity in context of the environment.

Ecology is distinguished from natural history, which deals primarily with the descriptive study of organisms. It is a sub-discipline of biology, which is the study of life.

There are many practical applications of ecology in conservation biology, wetland management, natural resource management (agriculture, forestry, fisheries), city planning (urban ecology), community health, economics, basic & applied science and it provides a conceptual framework for understanding and researching human social interaction (human ecology).

Levels of Organization and Study

Scale and Complexity

The scale and dynamics of time and space must be carefully considered when describing ecological phenomena. In reference to time, it can take thousands of years for ecological processes to mature. The life-span of a tree, for example, can include different successional or seral stages leading to mature old-growth forests. The ecological process is extended even further through time as trees topple over, decay and provide critical habitat as nurse logs or coarse woody debris. In reference to space, the area of an ecosystem can vary greatly from tiny to vast. For example, a single tree is of smaller consequence to the classification of a forest ecosystem, but it is of larger consequence to smaller organisms. Several generations of an aphid population, for example, might exist on a single leaf. Inside each of those aphids exist diverse communities of bacteria. Tree growth is, in turn, related to local site variables, such as soil type, moisture content, slope of the land, and forest canopy closure. However, more complex global factors, such as climate, must be considered for the classification and understanding of processes leading to larger patterns spanning across a forested landscape.

Global patterns of biological diversity are complex. This biocomplexity stems from the interplay among ecological processes that operate and influence patterns that grade into each other, such as transitional areas or ecotones that stretch across different scales. "Complexity in ecology is of at least six distinct types: spatial, temporal, structural, process, behavioral, and geometric." There are different views on what constitutes complexity. One perspective lumps things that we do not understand into this category by virtue of the computational effort it would require to piece together the numerous interacting parts. Alternatively, complexity in life sciences can be viewed as emergent self-organized systems with multiple possible outcomes directed by random accidents of history. Small scale patterns do not necessarily explain large scale phenomena, otherwise captured in the expression 'the sum is greater than the parts'. Ecologists have identified emergent and self-organizing phenomena that operate at different environmental scales of influence, ranging from molecular to planetary, and these require different sets of scientific explanation. Long-term ecological studies provide important track records to better understand the complexity of ecosystems over longer temporal and broader spatial scales. The International Long Term Ecological Network manages and exchanges scientific information among research sites. The longest experiment in existence is the Park Grass Experiment that was initiated in 1856. Another example includes the Hubbard Brook study in operation since 1960.

To structure the study of ecology into a manageable framework of understanding, the biological world is conceptually organized as a nested hierarchy of organization, ranging in scale from genes, to cells, to tissues, to organs, to organisms, to species and up to the level of the biosphere. Together these hierarchical scales of life form a panarchy. Ecosystems are primarily researched at (but not restricted to) three key levels of organization, including organisms, populations, and communities. Ecologists study ecosystems by sampling a certain number of individuals that are representative of a population. Ecosystems consist of communities interacting with each other and the environment. In ecology, communities are created by the interaction of the populations of different species in an area.

Biodiversity (an abbreviation of biological diversity) describes all varieties of life from genes to ecosystems and spans every level of biological organization. There are many ways to index, measure, and represent biodiversity. Biodiversity includes species diversity, ecosystem diversity, genetic diversity and the complex processes operating at and among these respective levels. Biodiversity plays an important role in ecological health as much as it does for human health. Preventing or prioritizing species extinctions is one way to preserve biodiversity, but populations, the genetic diversity within them and ecological processes, such as migration, are being threatened on global scales and disappearing rapidly as well. Conservation priorities and management techniques require different approaches and considerations to address the full ecological scope of biodiversity. Populations and species migration, for example, are more sensitive indicators of ecosystem services that sustain and contribute natural capital toward the well-being of humanity. An understanding of biodiversity has practical application for ecosystem-based conservation planners as they make ecologically responsible decisions in management recommendations to consultant firms, governments and industry.

Ecological Niche and the Habitat

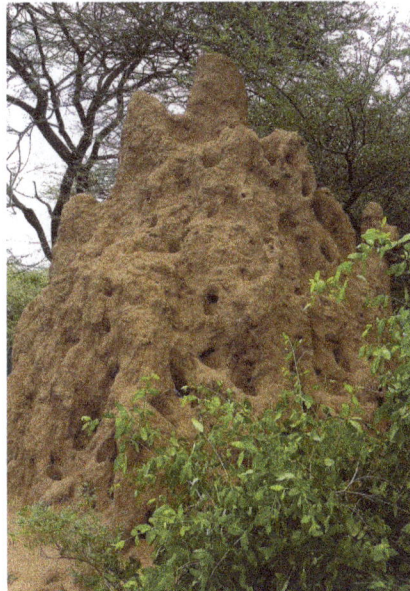

Termite mounds with varied heights of chimneys regulate gas exchange, temperature and other environmental parameters that are needed to sustain the internal physiology of the entire colony

There are many definitions of the niche dating back to 1917, but George Evelyn Hutchinson made conceptual advances in 1957 and introduced the most widely accepted definition: "The niche is the set of biotic and abiotic conditions in which a species is able to persist and maintain stable population sizes." The ecological niche is a central concept in the ecology of organisms and is sub-divided into the fundamental and the realized niche. The fundamental niche is the set of environmental conditions under which a species is able to persist. The realized niche is the set of environmental plus ecological conditions under which a species persists.

The habitat of a species is a related but distinct concept that describes the environment over which a species is known to occur and the type of community that is formed as a result. More specifically, "habitats can be defined as regions in environmental space that are composed of multiple dimensions, each representing a biotic or abiotic environmental variable; that is, any component or characteristic of the environment related directly (e.g. forage biomass and quality) or indirectly (e.g. elevation) to the use of a location by the animal. For example, the habitat might refer to an aquatic or terrestrial environment that can be further categorized as montane or alpine ecosystems.

Biogeographical patterns and range distributions are explained or predicted through knowledge and understanding of a species traits and niche requirements. Species have functional traits that are uniquely adapted to the ecological niche. A trait is a measurable property of an organism that influences its performance. Traits of each species are suited ar uniquely adapted to their ecological niche. This means that resident species are at an advantage and able to competitively exclude other similarly adapted species from having an overlapping geographic range. This is called the competitive exclusion principle.

Biodiversity of a coral reef. Corals adapt and modify their environment by forming calcium carbonate skeletons that provide growing conditions for future generations and form habitat for many other species

Organisms are subject to environmental pressures, but they are also modifiers of their habitats. The regulatory feedback between organisms and their environment can modify conditions from local (e.g., a pond) to global scales (e.g., Gaia), over time and even after death, such as decaying logs or silica skeleton deposits from marine organisms. The process and concept of ecosystem engineering has also been called niche construction. Ecosystem engineers are defined as: "...organisms that directly or indirectly modulate the availability of resources to other species, by causing physical state changes in biotic or abiotic materials. In so doing they modify, maintain and create habitats."

The ecosystem engineering concept has stimulated a new appreciation for the degree of influence that organisms have on the ecosystem and evolutionary process. The terms niche construction are more often used in reference to the under appreciated feedback mechanism of natural selection imparting forces on the abiotic niche. An example of natural selection through ecosystem engineering occurs in the nests of social insects, including ants, bees, wasps, and termites. There is an emergent homeostasis in the structure of the nest that regulates, maintains and defends the physiology of the entire colony. Termite mounds, for example, maintain a constant internal temperature through the design of air-conditioning chimneys. The structure of the nests themselves are subject to the forces of natural selection. Moreover, the nest can survive over successive generations, which means that ancestors inherit both genetic material and a legacy niche that was constructed before their time. Diatoms in the Bay of Fundy, Canada, provide another example of an ecosystem engineer. Benthic diatoms living in estuarine sediments secrete carbohydrate exudates that bind the sand and stabilizes the environment. The diatoms cause a physical state change in the properties of the sand that allows other organisms to colonize the area. The concept of ecosystem engineering brings new conceptual implications for the discipline of conservation biology.

Population Ecology

The population is the unit of analysis in population ecology. A population consists of individuals of the same species that live, interact and migrate through the same niche and habitat. A primary law of population ecology is the Malthusian growth model. This law states that:

"A population will grow (or decline) exponentially as long as the environment experienced by all individuals in the population remains constant."

This Malthusian premise provides the basis for formulating predictive theories and tests that follow. Simplified population models usually start with four variables including death, birth, immigration, and emigration. Mathematical models are used to calculate changes in population demographics using a null model. A null model is used as a null hypothesis for statistical testing. The null hypothesis states that random processes create observed patterns. Alternatively the patterns differ significantly from the random model and require further explanation. Models can be mathematically complex where "several competing hypotheses are simultaneously confronted with the data." An example of an introductory population model describes a closed population, such as on an island, where immigration and emigration does not take place. In these island models the per capita rates of change are described as:

$$dN/dT = B - D = bN - dN = (b-d)N = rN,$$

where N is the total number of individuals in the population, B is the number of births, D is the number of deaths, b and d are the per capita rates of birth and death respectively, and r is the per capita rate of population change. This formula can be read out as the rate of change in the population (dN/dT) is equal to births minus deaths (B − D).

Using these modelling techniques, Malthus' population principle of growth was later transformed into a model known as the logistic equation:

$$dN/dT = aN(1 - N/K),$$

where N is the number of individuals measured as biomass density, a is the maximum per-capita rate of change, and K is the carrying capacity of the population. The formula can be read as follows: the rate of change in the population (dN/dT) is equal to growth (aN) that is limited by carrying capacity (1 − N/K). The discipline of population ecology builds upon these introductory models to further understand demographic processes in real study populations and conduct statistical tests. The field of population ecology often uses data on life history and matrix algebra to develop projection matrices on fecundity and survivorship. This information is used for managing wildlife stocks and setting harvest quotas.

A list of terms that define various types of natural groupings of individuals that are used in population studies	
Term	Definition
Species population	All individuals of a species.
Metapopulation	A set of spatially disjunct populations, among which there is some immigration.
Population	A group of conspecific individuals that is demographically, genetically, or spatially disjunct from other groups of individuals.
Aggregation	A spatially clustered group of individuals.
Deme	A group of individuals more genetically similar to each other than to other individuals, usually with some degree of spatial isolation as well.
Local population	A group of individuals within an investigator-delimited area smaller than the geographic range of the species and often within a population (as defined above). A local population could be a disjunct population as well.
Subpopulation	An arbitrary spatially delimited subset of individuals from within a population (as defined above).

Metapopulation Ecology

Populations are also studied and modeled according to the metapopulation concept. The metapopulation concept was introduced in 1969: "as a population of populations which go extinct locally and recolonize." Metapopulation ecology is another statistical approach that is often used in conservation research. Metapopulation research simplifies the landscape into patches of varying levels of quality.

Metapopulation models have been used to explain life-history evolution, such as the ecological stability of amphibian metamorphosis in small vernal ponds. Alternative ecological strategies have evolved. For example, some salamanders forgo metamorphosis and sexually mature as aquatic neotenes. The seasonal duration of wetlands and the migratory range of the species determines which ponds are connected and if they form a metapopulation. The duration of the life history stages of amphibians relative to the duration of the vernal pool before it dries up regulates the ecological development of metapopulations connecting aquatic patches to terrestrial patches.

In metapopulation terminology there are emigrants (individuals that leave a patch), immigrants (individuals that move into a patch) and sites are classed either as sources or sinks. A site is a generic term that refers to places where ecologists sample populations, such as ponds or defined sampling areas in a forest. Source patches are productive sites that generate a seasonal supply of juveniles that migrate to other patch locations. Sink patches are unproductive sites that only receive migrants and will go extinct unless rescued by an adjacent source patch or environmental conditions become more favorable. Metapopulation models examine patch dynamics over time to answer questions about spatial and demographic ecology. The ecology of metapopulations is a dynamic process of extinction and colonization. Small patches of lower quality (i.e., sinks) are maintained or rescued by a seasonal influx of new immigrants. A dynamic metapopulation structure evolves from year to year, where some patches are sinks in dry years and become sources when conditions are more favorable. Ecologists use a mixture of computer models and field studies to explain metapopulation structure.

Ecosystem Ecology

The concept of the ecosystem was first introduced in 1935 to describe habitats within biomes that form an integrated whole and a dynamically responsive system having both physical and biological complexes. Within an ecosystem there are inseparable ties that link organisms to the physical and biological components of their environment to which they are adapted. Ecosystems are complex adaptive systems where the interaction of life processes form self-organizing patterns across different scales of time and space. This section introduces key areas of ecosystem ecology that are used to inquire, understand and explain oberved patterns of biodiversity and ecosystem function across different scales of organization.

Community Ecology

Community ecology is a subdiscipline of ecology which studies the distribution, abundance, demography, and interactions between coexisting populations. An example of a study in community ecology might measure primary production in a wetland in relation to decomposition and consumption rates. This requires an understanding of the community connections between plants (i.e., primary producers) and the decomposers (e.g., fungi and bacteria). or the analysis of predator-prey dynamics affecting amphibian biomass. Food webs and trophic levels are two widely employed conceptual models used to explain the linkages among species.

Food Webs

A food web is the archetypal ecological network. They are a type of concept map that illustrate pathways of energy flows in an ecological community, usually starting with solar energy being used by plants during photosynthesis. As plants grow, they accumulate carbohydrates and are eaten by grazing herbivores. Step by step lines or relations are drawn until a web of life is illustrated.

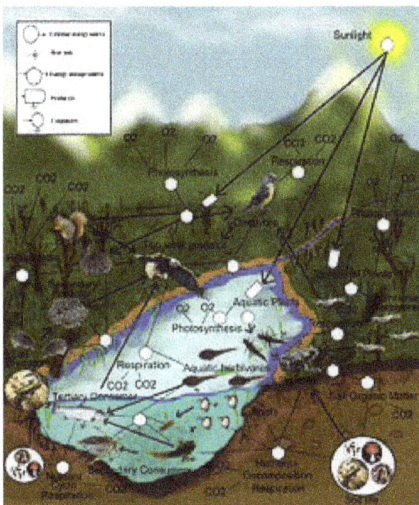

Freshwater aquatic and terrestrial food-webs

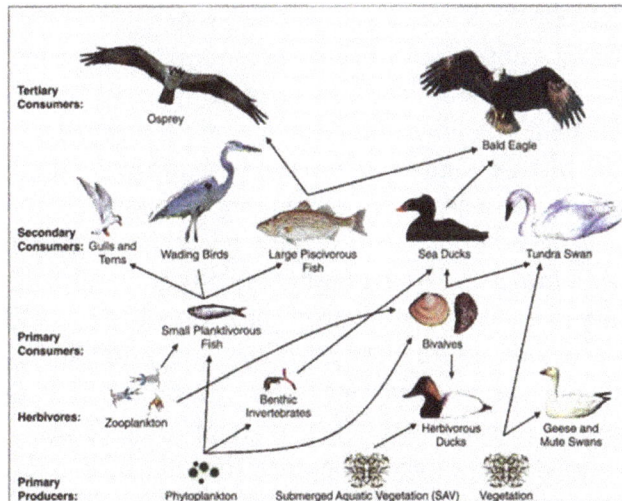

Generalized food web of waterbirds

There are different ecological dimensions that can be mapped to create more complicated food webs, including: species composition (type of species), richness (number of species), biomass (the dry weight of plants and animals), productivity (rates of conversion of energy and nutrients into growth), and stability (food webs over time). A food web diagram illustrating species composition shows how change in a single species can directly and indirectly influence many others. Microcosm studies are used to simplify food web research into semi-isolated units such as small springs, decaying logs, and laboratory experiments using organisms that reproduce quickly, such as daphnia feeding on algae grown under controlled environments in jars of water.

Principles gleaned from food web microcosm studies are used to extrapolate smaller dynamic concepts to larger systems. Food webs are limited because they are generally restricted to a specific habitat, such as a cave or a pond. The food web illustration (right) only shows a small part of the complexity connecting the aquatic system to the adjacent terrestrial land. Many of these species migrate into other habitats to distribute their effects on a larger scale. In other words, food webs are incomplete, but are nonetheless a valuable tool in understanding community ecosystems.

Food chain length is another way of describing food webs as a measure of the number of species encountered as energy or nutrients move from the plants to top predators. There are different ways of calculating food chain length depending on what parameters of the food web dynamic are being considered: connectance, energy, or interaction. In a simple predator-prey example, a deer is one step removed from the plants it eats (chain length = 1) and a wolf that eats the deer is two steps removed (chain length = 2). The relative amount or strength of influence that these parameters have on the food web address questions about:

- the identity or existence of a few dominant species (called strong interactors or keystone species)

- the total number of species and food-chain length (including many weak interactors) and

- how community structure, function and stability is determined.

Trophic Dynamics

The Greek root of the word troph, means food or feeding. Links in food-webs primarily connect feeding relations or trophism among species. Biodiversity within ecosystems can be organized into vertical and horizontal dimensions. The vertical dimension represents feeding relations that become further removed from the base of the food chain up toward top predators. The horizontal dimension represents the abundance or biomass at each level. When the relative abundance or biomass of each functional feeding group is stacked into their respective trophic levels they naturally sort into a 'pyramid of numbers'. Functional groups are broadly categorized as autotrophs (or primary producers), heterotrophs (or consumers), and detrivores (or decomposers). Heterotrophs can be further sub-divided into different functional groups, including: primary consumers (strict herbivores), secondary consumers (predators that feed exclusively on herbivores) and tertiary consumers (predators that feed on a mix of herbivores and predators). Omnivores do not fit neatly into a functional category because they eat both plant and animal tissues. It has been suggested, however, that omnivores have a greater functional influence as predators because relative to herbivores they are comparatively inefficient at grazing.

Ecologist collect data on trophic levels and food webs to statistically model and mathematically calculate parameters, such as those used in other kinds of network analysis (e.g., graph theory), to study emergent patterns and properties shared among ecosystems. The emergent pyramidal arrangement of trophic levels with amounts of energy transfer decreasing as species become further removed from the source of production is one of several patterns that is repeated amongst the planets ecosystems. The size of each level in the pyramid generally represents biomass, which can be measured as the dry weight of an organism. Autotrophs may have the highest global proportion of biomass, but they are closely rivaled or surpassed by microbes.

The decomposition of dead organic matter, such as leaves falling on the forest floor, turns into soils that feed plant production. The total sum of the planet's soil ecosystems is called the pedosphere where a very large proportion of the Earth's biodiversity sorts into other trophic levels. Invertebrates that feed and shred larger leaves, for example, create smaller bits for smaller organisms in the feeding chain. Collectively, these are the detrivores that regulate soil formation. Tree roots, fungi, bacteria, worms, ants, beetles, centipedes, spiders, mammals, birds, reptiles, amphibians and other less familiar creatures all work to create the trophic web of life in soil ecosystems. As organisms feed and migrate through soils they physically displace materials, which is an important ecological process called bioturbation. Biomass of soil microorganisms are influenced by and feed back into the trophic dynamics of the exposed solar surface ecology. Paleoecological studies of soils places the origin for bioturbation to a time before the Cambrian period. Other events, such as the evolution of trees and amphibians moving into land in the Devonian period played a significant role in the development of soils and ecological trophism.

Functional trophic groups sort out hierarchically into pyramidic trophic levels because it requires specialized adaptations to become a photosynthesizer or a predator, so few organisms have the adaptations needed to combine both abilities. This explains why functional adaptations to trophism (feeding) organizes different species into emergent functional groups. Trophic levels are part of the holistic or complex systems view of ecosystems. Each trophic level contains unrelated species that grouped together because they share common ecological functions. Grouping functionally similar species into a trophic system gives a macroscopic image of the larger functional design.

Links in a food-web illustrate direct trophic relations among species, but there are also indirect effects that can alter the abundance, distribution, or biomass in the trophic levels. For example, predators eating herbivores indirectly influence the control and regulation of primary production in plants. Although the predators do not eat the plants directly, they regulate the population of herbivores that are directly linked to plant trophism. The net effect of direct and indirect relations is called trophic cascades. Trophic cascades are separated into species-level cascades, where only a subset of the food-web dynamic is impacted by a change in population numbers, and community-level cascades, where a change in population numbers has a dramatic effect on the entire food-web, such as the distribution of plant biomass.

Keystone Species

A keystone species is a species that is disproportionately connected to more species in the food-web. Keystone species have lower levels of biomass in the trophic pyramid relative to the importance of their role. The many connections that a keystone species holds means that it maintains the organization and structure of entire communities. The loss of a keystone species results in a range of dramatic cascading effects that alters trophic dynamics, other food-web connections and can cause the extinction of other species in the community.

Sea otters (Enhydra lutris) are commonly cited as an example of a keystone species because they limit the density of sea urchins that feed on kelp. If sea otters are removed from the system, the urchins graze until the kelp beds disappear and this has a dramatic effect on community structure. Hunting of sea otters, for example, is thought to have indirectly led to the extinction of the Steller's Sea Cow (Hydrodamalis gigas). While the keystone species concept has been used extensively as a conservation tool, it has been criticized for being poorly defined from an operational stance. It is very difficult to experimentally determine in each different ecosystem what species may hold a keystone role. Furthermore, food-web theory suggests that keystone species may not be all that common. It is therefore unclear how generally the keystone species model can be applied.

The Biome

Ecological units of organization are defined through reference to any magnitude of space and time on the planet. Communities of organisms, for example, are somewhat arbitrarily defined, but the processes of life integrate at different levels and organize into more complex wholes. Biomes, for example, are a larger unit of organization that categorize regions of the Earth's ecosystems main-ly according to the structure and composition of vegetation. Different researchers have applied different methods to define continental boundaries of biomes dominated by different functional types of vegetative communities that are limited in distribution by climate, precipitation, weather and other environmental variables. Examples of biome names include: tropical rainforest, tem-perate broadleaf and mixed forests, temperate deciduous forest, taiga, tundra, hot desert, and polar desert. Other researchers have recently started to categorize other types of biomes, such as the human and oceanic microbiomes. To a microbe, the human body is a habitat and a landscape. The microbiome has been largely discovered through advances in molecular genetics that have revealed a hidden richness of microbial diversity on the planet. The oceanic microbiome plays a significant role in the ecological biogeochemistry of the planet's oceans.

The Biosphere

Ecological theory has been used to explain self-emergent regulatory phenomena at the planetary scale. The largest scale of ecological organization is the biosphere: the total sum of ecosystems on the planet. Ecological relations regulate the flux of energy, nutrients, and climate all the way up to the planetary scale. For example, the dynamic history of the planetary CO_2 and O_2 composition of the atmosphere has been largely determined by the biogenic flux of gases coming from respiration and photosynthesis, with levels fluctuating over time and in relation to the ecology and evolution of plants and animals. When sub-component parts are organized into a whole there are oftentimes emergent properties that describe the nature of the system. This the Gaia hypothesis, and is an example of holism applied in ecological theory. The ecology of the planet acts as a single regulato-ry or holistic unit called Gaia. The Gaia hypothesis states that there is an emergent feedback loop generated by the metabolism of living organisms that maintains the temperature of the Earth and atmospheric conditions within a narrow self-regulating range of tolerance.

Relation to Evolution

Ecology and evolution are considered sister disciplines of the life sciences. Natural selection, life history, development, adaptation, populations, and inheritance are examples of concepts that thread equally into ecological and evolutionary theory. Morphological, behavioural and/ or genetic traits, for example, can be mapped onto evolutionary trees to study the historical development of a species in relation to their functions and roles in different ecological circumstances. In this framework, the analytical tools of ecologists and evolutionists overlap as they organize, classify and investigate life through common systematic principals, such as phylogenetics or the Linnaean system of taxonomy. The two disciplines often appear together, such as in the title of the journal Trends in Ecology and Evolution. There is no sharp boundary separating ecology from evolution and they differ more in their areas of applied focus. Both disciplines discover and explain emergent and unique properties and processes operating across different spatial or temporal scales of organization. While the boundary between ecology and evolution is not always clear, it is understood that ecologists study the abiotic and biotic factors that influence the evolutionary process.

Behavioral Ecology

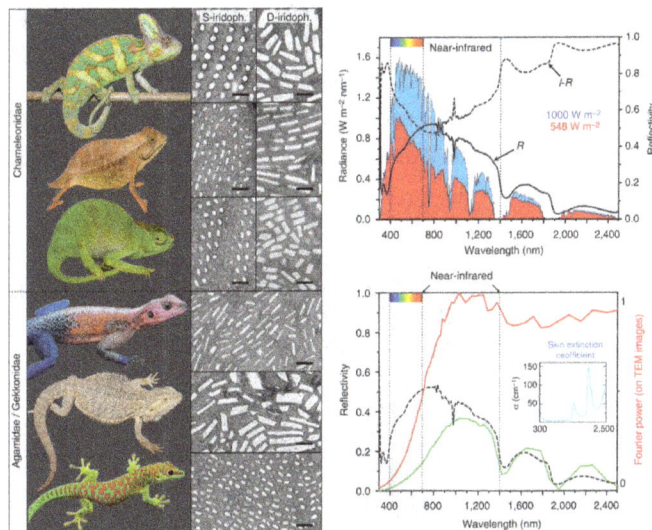

Social display and color variation in differently adapted species of chameleons (Bradypodion spp.). Chameleons change their skin color to match their background as a behavioral defense mechanism and also use color to communicate with other members of their species

All organisms are motile to some extent. Even plants express complex behavior, including memory and communication. Behavioural ecology is the study of ethology and its ecological and evolutionary implications. Ethology is the study of observable movement or behaviour in nature. This could include investigations of motile sperm of plants, mobile phytoplankton, zooplankton swimming toward the female egg, the cultivation of fungi by weevils, the mating dance of a salamander, or social gatherings of amoeba.

Adaptation is the central unifying concept in behavioral ecology."International Society for Behavioral Ecology". http://www.behavecol.com/pages/society/welcome.html.Behaviors can be recorded as traits and inherited in much the same way that eye and hair color can. Behaviours evolve and become adapted to the ecosystem because they are subject to the forces of natural selection. Hence, behaviors can be adaptive, meaning that they evolve functional utilities that increases reproductive success for the individuals that inherit such traits. This is also the technical definition for fitness in biology, which is a measure of reproductive success over successive generations.

Predator-prey interactions are an introductory concept into food-web studies as well as behavioural ecology. Prey species can exhibit different kinds of behavioural adaptations to predators, such as avoid, flee or defend. Many prey species are faced with multiple predators that differ in the degree of danger posed. To be adapted to their environment and face predatory threats, organisms must balance their energy budgets as they invest in different aspects of their life history, such as growth, feeding, mating, socializing, or modifying their habitat. Hypotheses posited in behavioural ecology are generally based on adaptive principals of conservation, optimization or efficiency. For example,

"The threat-sensitive predator avoidance hypothesis predicts that prey should assess the degree of threat posed by different predators and match their behavior according to current levels of risk."

"The optimal flight initiation distance occurs where expected postencounter fitness is maximized, which depends on the prey's initial fitness, benefits obtainable by not fleeing, energetic escape costs, and expected fitness loss due to predation risk."

The behaviour of long-toed salamanders (Ambystoma macrodactylum) presents an example in this context. When threatened, the long-toed salamander defends itself by waving its tail and secreting a white milky fluid. The excreted fluid is distasteful, toxic and adhesive, but it is also used for nutrient and energy storage during hibernation. Hence, salamanders subjected to frequent predatory attack will be energetically compromised as they use up their energy stores.

Symbiosis: Leafhoppers (Eurymela fenestrata) are protected by ants (Iridomyrmex purpureus) in a symbiotic relationship. The ants protect the leafhoppers from predators and in return the leafhoppers feeding on plants exude honeydew from their anus that provides energy and nutrients to tending ants

Ecological interactions can be divided into host and associate relationships. A host is any entity that harbors another that is called the associate. Host and associate relationships among species that are mutually or reciprocally beneficial are called mutualisms. If the host and associate are physically connected, the relationship is called symbiosis. Approximately 60% of all plants, for example, have a symbiotic relationship with arbuscular mycorrhizal fungi. Symbiotic plants and fungi exchange carbohydrates for mineral nutrients. Symbiosis differs from indirect mutualisms where the organisms live apart. For example, tropical rainforests regulate the Earth's atmosphere. Trees living in the equatorial regions of the planet supply oxygen into the atmosphere that sustains species living in distant polar regions of the planet. This relationship is called commensalism because many other host species receive the benefits of clean air at no cost or harm to the associate tree species supplying the oxygen. The host and associate relationship is called parasitism if one species benefits while the other suffers. Competition among species or among members of the same species is defined as reciprocal antagonism, such as grasses competing for growth space.

Popular ecological study systems for mutualism include, fungus-growing ants employing agricultural symbiosis, bacteria living in the guts of insects and other organisms, the fig wasp and yucca moth pollination complex, lichens with fungi and photosynthetic algae, and corals with photosynthetic algae.

Intraspecific behaviours are notable in the social insects, slime moulds, social spiders, human society, and naked mole rats where eusocialism has evolved. Social behaviours include reciprocally beneficial behaviours among kin and nest mates. Social behaviours evolve from kin and group selection. Kin selection explains altruism through genetic relationships, whereby an altruistic behaviour leading to death is rewarded by the survival of genetic copies distributed among surviving relatives. The social insects, including ants, bees and wasps are most famously studied for this type of relationship because the male drones are clones that share the same genetic make-up as every other male in the colony. In contrast, group selectionists find examples of altruism among non-genetic relatives and explain this through selection acting on the group, whereby it becomes selectively advantageous for groups if their members express altruistic behaviours to one another. Groups that are predominantely altruists beat groups that are predominantely selfish.

A often quoted behavioural ecology hypothesis is known as Lack's brood reduction hypothesis (named after David Lack). Lack's hypothesis posits an evolutionary and ecological explanation as to why birds lay a series of eggs with an asynchronous delay leading to nestlings of mixed age and weights. According to Lack, this brood behaviour is an ecological insurance that allows the larger birds to survive in poor years and all birds to survive when food is plentiful.

Elaborate sexual displays and posturing are encountered in the behavioural ecology of animals. The birds of paradise, for example, display elaborate ornaments and song during courtship. These displays serve a dual purpose of signalling healthy or well-adapted individuals and good genes. The elaborate displays are driven by sexual selection as an advertisement of quality of traits among male suitors.

Biogeography

The word biogeography is an amalgamation of biology and geography. Biogeography is the comparative study of the geographic distribution of organisms and the corresponding evolution of their traits in space and time. The Journal of Biogeography was established in 1974. Biogeography and ecology share many of their disciplinary roots. For example, the theory of island biogeography, published by the mathematician Robert MacArthur and ecologist Edward O. Wilson in 1967 is considered one of the fundamentals of ecological theory.

Biogeography has a long history in the natural sciences where questions arise concerning the spatial distribution of plants and animals. Ecology and evolution provide the explanatory context for biogeographical studies. Biogeographical patterns result from ecological processes that influence range distributions, such as migration and dispersal. and from historical processes that split populations or species into different areas. The biogeographic processes that result in the natural splitting of species explains much of the modern distribution of the Earth's biota. The splitting of lineages in a species is called vicariance biogeography and it is a sub-discipline of biogeography. There are also practical applications in the field of biogeography concerning ecological systems and processes. For example, the range and distribution of biodiversity and invasive species responding to climate change is a serious concern and active area of research in context of global warming.

R/K-Selection Theory

Another concept that was introduced in MacArthur and Wilson's (1967) classical book, The Theory of Island Biogeography was r/K selection theory, which was one of the first predictive models in ecology that could be used to explain life-history evolution. The premise behind the r/K selection model is that the pressure of natural selection changes according to population densities. When an island is first colonized the density of individuals is low. The initial increase in population size is not limited by competition, which leaves an abundance of available resources for rapid population growth. These early phases of population growth experience density independent forces of natural selection, which is called r-selection. When the population becomes crowded, it reaches the island's carrying capacity, and individuals compete more heavily for fewer available resources. Under crowded conditions the population experiences density-dependent forces of natural selection, called K-selection.

In the r/K-selection model, the first variable r is the intrinsic rate of natural increase in population size and the second variable K is the carrying capacity of a population. Different species evolve different life-history strategies spanning a continuum between these two selective forces. An r-selected species is one that has high birth rates, low levels of parental investment, and high rates of mortality before individuals reach maturity. Evolution favors high rates of fecundity in r-selected species. Many kinds of insects and invasive species exhibit r-selected characteristics. In contrast, a K-selected species has low rates of fecundity, high levels of parental investment in the young, and low rates of mortality as individuals mature. Humans and elephants are examples of species exhibiting K-selected characteristics, including longevity and efficiency in the conversion of more resources into fewer offspring.

Molecular Ecology

The important relationship between ecology and genetic inheritance predates modern techniques for molecular analysis. Molecular ecological research became more feasible with the development of rapid and accessible genetic technologies, such as the polymerase chain reaction (PCR). The rise of molecular technologies and influx of research questions into this new ecological field resulted in the publication Molecular Ecology in 1992. Molecular ecology uses various analytical techniques to study genes in an evolutionary and ecological context. In 1994, professor John Avise also played a leading role in this area of science with the publication of his book, Molecular Markers, Natural History and Evolution. Newer technologies opened a wave of genetic analysis into organisms once difficult to study from an ecological or evolutionary standpoint, such as bacteria, fungi and nematodes.

Molecular ecology engendered a new research paradigm to investigate ecological questions considered otherwise intractable. Molecular investigations revealed previously obscured details in the tiny intricacies of nature and improved resolution into probing questions about behavioral and biogeographical ecology. For example, molecular ecology revealed promiscuous sexual behavior and multiple male partners in tree swallows previously thought to be socially monogamous. In a biogeographical context, the marriage between genetics, ecology and evolution resulted in a new sub-discipline called phylogeography.

Relation to the Environment

The environment is dynamically interlinked, imposed upon and constrains organisms at any time throughout their life cycle. Like the term ecology, environment has different conceptual meanings and to many these terms also overlap with the concept of nature. Environment "... includes the physical world, the social world of human relations and the built world of human creation." The environment in ecosystems includes both physical parameters and biotic attributes. The physical environment is external to the level of biological organization under investigation, including abiotic factors such as temperature, radiation, light, chemistry, climate and geology. The biotic environment includes genes, cells, organisms, members of the same species (conspecifics) and other species that share a habitat. The laws of thermodynamics applies to ecology by means of its physical state. Armed with an understanding of metabolic and thermodynamic principles a complete accounting of energy and material flow can be traced through an ecosystem.

Environmental and ecological relations are studied through reference to conceptually manageable and isolated parts. However, once the effective environmental components are understood they conceptually link back together as a holocoenotic system. In other words, the organism and the environment form a dynamic whole (or umwelt). Change in one ecological or environmental factor can concurrently affect the dynamic state of an entire ecosystem.

Ecological studies are necessarily holistic as opposed to reductionistic. Holism has three scientific meanings or uses that identify with: 1) the mechanistic complexity of ecosystems, 2) the practical description of patterns in quantitative reductionist terms where correlations may be identified but nothing is understood about the causal relations without reference to the whole system, which leads to 3) a metaphysical hierarchy whereby the causal relations of larger systems are understood without reference to the smaller parts. An example of the metaphysical aspect to holism is the trend of increased exterior thickness in shells of different species. The reason for a thickness increase can be understood through reference to principals of natural selection via predation without any reference to the biomolecular properties of the exterior shells.

Metabolism and the Early Atmosphere

The Earth formed approximately 4.5 billion years ago and environmental conditions were too extreme for life to form for the first 500 million years. During this early Hadean period, the Earth started to cool, allowing a crust and oceans to form. Environmental conditions were unsuitable for the origins of life for the first billion years after the Earth formed. The Earth's atmosphere transformed from being dominated by hydrogen, to one composed mostly of methane and ammonia. Over the next billion years the metabolic activity of life transformed the atmosphere to higher concentrations of carbon dioxide, nitrogen, and water vapor. These gases changed the way that light from the sun hit the Earth's surface and greenhouse effects trapped heat. There were untapped sources of free energy within the mixture of reducing and oxidizing gasses that set the stage for primitive ecosystems to evolve and, in turn, the atmosphere also evolved.

The leaf is the primary site of photosynthesis in most plants

Throughout history, the Earth's atmosphere and biogeochemical cycles have been in a dynamic equilibrium with planetary ecosystems. The history is characterized by periods of significant transformation followed by millions of years of stability. The evolution of the earliest organisms, likely anaerobic methanogen microbes, started the process by converting atmospheric hydrogen into methane ($4H_2 + CO_2 \rightarrow CH_4 + 2H_2O$). Anoxygenic photosynthesis converting hydrogen sulfide into other sulfur compounds or water ($2H_2S + CO_2 \rightarrow hv \rightarrow CH_2O \rightarrow H_2O \rightarrow + 2S$ or $2H_2 + CO_2 + hv \rightarrow CH_2O + H_2O$), as occurs in deep sea hydrothermal vents today, reduced hydrogen concentrations and increased atmospheric methane. Early forms of fermentation also increased levels of atmospheric methane. The transition to an oxygen dominant atmosphere (the Great Oxidation) did not begin until approximately 2.4-2.3 billion years ago, but photosynthetic processes started 0.3 to 1 billion years prior.

Radiation: Heat, Temperature and Light

The biology of life operates within a certain range of temperatures. Heat is a form of energy that regulates temperature. Heat affects growth rates, activity, behavior and primary production. Temperature is largely dependent on the incidence of solar radiation. The latitudinal and longitudinal spatial variation of temperature greatly affects climates and consequently the distribution of biodiversity and levels of primary production in different ecosystems or biomes across the planet. Heat and temperature relate importantly to metabolic activity. Poikilotherms, for example, have a body temperature that is largely regulated and dependent on the temperature of the external environment. In contrast, homeotherms regulate their internal body temperature by expending metabolic energy.

There is a relationship between light, primary production, and ecological energy budgets. Sunlight is the primary input of energy into the planet's ecosystems. Light is composed of electromagnetic energy of different wavelengths. Radiant energy from the sun generates heat, provides photons of light measured as active energy in the chemical reactions of life, and also acts as a catalyst for genetic mutation. Plants, algae, and some bacteria absorb light and assimilate the energy through photosynthesis. Organisms capable of assimilating energy by photosynthesis or through inorganic fixation of H_2S are autotrophs. Autotrophs—responsible for primary production—assimilate light energy that becomes metabolically stored as potential energy in the form of biochemical enthalpic bonds.

Physical Environments

Water

The rate of diffusion of carbon dioxide and oxygen is approximately 10,000 times slower in water than it is in air. When soils become flooded, they quickly lose oxygen from low-concentration (hypoxic) to an (anoxic) environment where anaerobic bacteria thrive among the roots. Water also influences the spectral properties of light that becomes more diffuse as it is reflected off the water surface and submerged particles. Aquatic plants exhibit a wide variety of morphological and physiological adaptations that allow them to survive, compete and diversify these environments. For example, the roots and stems develop large cellular air spaces to allow for the efficient transportation gases (for example, CO_2 and O_2) used in respiration and photosynthesis. In drained soil, microorganisms use oxygen during respiration. In aquatic environments, anaerobic soil microorganisms use nitrate, manganic ions, ferric ions, sulfate, carbon dioxide and some organic compounds. The activity of soil microorganisms and the chemistry of the water reduces the oxidation-reduction potentials of the water. Carbon dioxide, for example, is reduced to methane (CH_4) by methanogenic bacteria. Salt water also requires special physiological adaptations to deal with water loss. Salt water plants (or halophytes) are able to osmo-regulate their internal salt (NaCl) concentrations or develop special organs for shedding salt away. The physiology of fish is also specially adapted to deal with high levels of salt through osmoregulation. Their gills form electrochemical gradients that mediate salt excrusion in salt water and uptake in fresh water.

Gravity

The shape and energy of the land is affected to a large degree by gravitational forces. On a larger scale, the distribution of gravitational forces on the earth are uneven and influence the shape and movement of tectonic plates as well as having an influence on geomorphic processes such as orogeny and erosion. These forces govern many of the geophysical properties and distributions of ecological biomes across the Earth. On a organism scale, gravitational forces provide directional cues for plant and fungal growth (gravitropism), orientation cues for animal migrations, and influence the biomechanics and size of animals. Ecological traits, such as allocation of biomass in trees during growth are subject to mechanical failure as gravitational forces influence the position and structure of branches and leaves. The cardiovascular systems of all animals are functionally adapted to overcome pressure and gravitational forces that change according to the features of organisms (e.g., height, size, shape), their behavior (e.g., diving, running, flying), and the habitat occupied (e.g., water, hot deserts, cold tundra).

Pressure

Climatic and osmotic pressure places physiological constraints on organisms, such as flight and respiration at high altitudes, or diving to deep ocean depths. These constraints influence vertical limits of ecosystems in the biosphere as organisms are physiologically sensitive and adapted to atmospheric and osmotic water pressure differences. Oxygen levels, for example, decrease with increasing pressure and are a limiting factor for life at higher altitudes. Water transportation through trees is another important ecophysiological parameter dependent upon pressure. Water pressure in the depths of oceans requires adaptations to deal with the different living conditions. Mammals, such as whales, dolphins and seals are adapted to deal with changes in sound due to water pressure differences.

Wind and Turbulence

Turbulent forces in air and water have significant effects on the environment and ecosystem distribution, form and dynamics. On a planetary scale, ecosystems are affected by circulation patterns in the global trade winds. Wind power and the turbulent forces it creates can influence heat, nutrient, and biochemical profiles of ecosystems. For example, wind running over the surface of a lake creates turbulence, mixing the water column and influencing the environmental profile to create thermally layered zones, partially governing how fish, algae, and other parts of the aquatic ecology are structured. Wind speed and turbulence also exert influence on rates of evapotranspiration rates and energy budgets in plants and animals. Wind speed, temperature and moisture content can vary as winds travel across different landfeatures and elevations. The westerlies, for example, come into contact with the coastal and interior mountains of western North America to produce a rain shadow on the leeward side of the mountain. The air expands and moisture condenses as the winds move up in elevation which can cause precipitation; this is called orographic lift. This environmental process produces spatial divisions in biodiversity, as species adapted to wetter conditions are range-restricted to the coastal mountain valleys and unable to migrate across the xeric ecosystems of the Columbia Basin to intermix with sister lineages that are segregated to the interior mountain systems.

Fire

Forest fires modify the land by leaving behind an environmental mosaic that diversifies the landscape into different seral stages and habitats of varied quality (left). Some species are adapted to forest fires, such as pine trees that open their cones only after fire exposure (right)

Plants convert carbon dioxide into biomass and emit oxygen into the atmosphere. Approximately 350 million years ago (near the Devonian period) the photosynthetic process brought the concentration of atmospheric oxygen above 17%, which allowed combustion to occur. Fire releases CO_2 and converts fuel into ash and tar. Fire is a significant ecological parameter that raises many issues pertaining to its control and suppression in management. While the issue of fire in relation to ecology and plants has been recognized for a long time, Charles Cooper brought attention to the issue of forest fires in relation to the ecology of forest fire suppression and management in the 1960s.

Fire creates environmental mosaics and a patchiness to ecosystem age and canopy structure.

Native North Americans were among the first to influence fire regimes by controlling their spread near their homes or by lighting fires to stimulate the production of herbaceous foods and basketry materials. The altered state of soil nutrient supply and cleared canopy structure also opens new ecological niches for seedling establishment. Most ecosystem are adapted to natural fire cycles. Plants, for example, are equipped with a variety of adaptations to deal with forest fires. Some species (e.g., Pinus halepensis) cannot germinate until after their seeds have lived through a fire. This environmental trigger for seedlings is called serotiny. Some compounds from smoke also promote seed germination.

Biogeochemistry

Ecologists study and measure nutrient budgets to understand how these materials are regulated and flow through the environment. This research has led to an understanding that there is a global feedback between ecosystems and the physical parameters of this planet including minerals, soil, pH, ions, water and atmospheric gases. There are six major elements, including H (hydrogen), C (carbon), N (nitrogen), O (oxygen), S (sulfur), and P (phosphorus) that form the constitution of all biological macromolecules and feed into the Earth's geochemical processes. From the smallest scale of biology the combined effect of billions upon billions of ecological processes amplify and ultimately regulate the biogeochemical cycles of the Earth. Understanding the relations and cycles mediated between these elements and their ecological pathways has significant bearing toward understanding global biogeochemistry.

The ecology of global carbon budgets gives one example of the linkage between biodiversity and biogeochemistry. For starters, the Earth's oceans are estimated to hold 40,000 gigatonnes (Gt) carbon, vegetation and soil is estimated to hold 2070 Gt carbon, and fossil fuel emissions are estimated to emit an annual flux of 6.3 Gt carbon. At different times in the Earth's history there has been major restructuring in these global carbon budgets that was regulated to a large extent by the ecology of the land. For example, through the early-mid Eocene volcanic outgassing, the oxidation of methane stored in wetlands, and seafloor gases increased atmospheric CO_2 concentrations to levels as high as 3500 ppm. In the Oligocene, from 25 to 32 million years ago, there was another significant restructuring in the global carbon cycle as grasses evolved a special type of C4 photosynthesis and expanded their ranges. This new photosynthetic pathway evolved in response to the drop in atmospheric CO_2 concentrations below 550 ppm. Ecosystem functions such as these feed back significantly into global atmospheric models for carbon cycling. Loss in the abundance and distribution of biodiversity causes global carbon cycle feedbacks that are expected to increase rates of global warming in the next century. The effect of global warming melting large sections of permafrost creates a new mosaic of flooded areas where decomposition results in the emission of methane (CH_4). Hence, there is a relationship between global warming, decomposition and respiration in soils and wetlands producing significant climate feedbacks and altered global biogeochemical cycles. There is concern over increases in atmospheric methane in the context of the global carbon cycle, because methane is also a greenhouse gas that is 23 times more effective at absorbing long-wave radiation on a 100 year time scale.

Ecosystem Services and the Biodiversity Crisis

The ecosystems of planet Earth are coupled to human environments. Ecosystems regulate the global geophysical cycles of energy, climate, soil nutrients, and water that in turn support and grow natural capital (including the environmental, physiological, cognitive, cultural, and spiritual dimensions of life). Ultimately, every manufactured product in human environments comes from natural systems. Ecosystems are considered common-pool resources because ecosystems do not exclude beneficiaries and they can be depleted or degraded. For example, green space within communities provides common-pool health services. Research shows that people who are more engaged with regular access to natural areas have lower rates of diabetes, heart disease and psychological disorders. These ecological health services are regularly depleted through urban development projects that do not factor in the common-pool value of ecosystems.

A bumblebee pollinating a flower, one example of an ecosystem service

The ecological commons delivers a diverse supply of community services that sustains the well-being of human society. The Millennium Ecosystem Assessment, an international UN initiative involving more than 1,360 experts worldwide, identifies four main ecosystem service types having 30 sub-categories stemming from natural capital. The ecological commons includes provisioning (e.g., food, raw materials, medicine, water supplies), regulating (e.g., climate, water, soil retention, flood retention), cultural (e.g., science and education, artistic, spiritual), and supporting (e.g., soil formation, nutrient cycling, water cycling) services.

Ecological economics is an economic science that uses many of the same terms and methods that are used in accounting. Natural capital is the stock of materials or information stored in biodiversity that generates services that can enhance the welfare of communities. Population losses are the more sensitive indicator of natural capital than are species extinction in the accounting of ecosystem services. The prospect for recovery in the economic crisis of nature is grim. Populations, such as local ponds and patches of forest are being cleared away and lost at rates that exceed species extinctions.

The WWF 2008 living planet report and other researchers report that human civilization has exceeded the bio-regenerative capacity of the planet. This means that human consumption is extracting more natural resources than can be replenished by ecosystems around the world. In 1992, professor William Rees developed the concept of our ecological footprint. The ecological footprint is a way of accounting the level of impact that human development is having on the Earth's ecosystems. All indications are that the human enterprise is unsustainable as the ecological footprint of society is placing too much stress on the ecology of the planet. The mainstream growth-based economic system adopted by governments worldwide does not include a price or markets for natural capital. This type of economic system places further ecological debt onto future generations.

Human societies are increasingly being placed under stress as the ecological commons is diminished through an accounting system that has incorrectly assumed "... that nature is a fixed, indestructible capital asset." While nature is resilient and it does regenerate, there are limits to what can be extracted, but conventional monetary analyses are unable to detect the problem. Evidence of the limits in natural capital are found in the global assessments of biodiversity, which indicate that the current epoch, the Anthropocene is a sixth mass extinction. Species loss is accelerating at 100–1000 times faster than average background rates in the fossil record. The ecology of the planet has been radically transformed by human society and development causing massive loss of ecosystem services that otherwise deliver and freely sustain equitable benefits to human society through the ecological commons. The ecology of the planet is further threatened by global warming, but investments in nature conservation can provide a regulatory feedback to store and regulate carbon and other greenhouse gases. The field of conservation biology involves ecologists that are researching the nature of the biodiversity threat and searching for solutions to sustain the planet's ecosystems for future generations.

"Human activities are associated directly or indirectly with nearly every aspect of the current extinction spasm."

The current wave of threats, including massive extinction rates and concurrent loss of natural capital to the detriment of human society, is happening rapidly. This is called a biodiversity crisis, because 50% of the worlds species are predicted to go extinct within the next 50 years. The world's fisheries are facing dire challenges as the threat of global collapse appears imminent, with serious ramifications for the well-being of humanity. Governments of the G8 met in 2007 and set forth 'The Economics of Ecosystems and Biodiversity' (TEEB) initiative :

In a global study we will initiate the process of analyzing the global economic benefit of biological diversity, the costs of the loss of biodiversity and the failure to take protective measures versus the costs of effective conservation.

Ecologists are teaming up with economists to measure the wealth of ecosystems and to express their value as a way of finding solutions to the biodiversity crisis. Some researchers have attempted to place a dollar figure on ecosystem services, such as the value that the Canadian boreal forest is contributing to global ecosystem services. If ecologically intact, the boreal forest has an estimated value of US\$3.7 trillion. The boreal forest ecosystem is one of the planet's great atmospheric regulators and it stores more carbon than any other biome on the planet. The annual value for ecological services of the Boreal Forest is estimated at US\$93.2 billion, or 2.5 greater than the annual value of resource extraction. The economic value of 17 ecosystem services for the entire biosphere (calculated in 1997) has an estimated average value of US\$33 trillion (10^{12}) per year. These ecological economic values are not currently included in calculations of national income accounts, the GDP and they have no price attributes because they exist mostly outside of the global markets. The loss of natural capital continues to accelerate and goes undetected by mainstream monetary analysis.

Forest Ecology

Forest ecology is the study of forest ecosystems. Forests are ecosystems in which the major ecological characteristics reflect the dominance of ecosystem conditions and processes by trees. Ecosystems are ecological systems that have the attributes of structure, function, interaction of the component parts, complexity (that reflects the structure, function and interactions) and change over time. An ecosystem can be of almost any physical size as long as it exhibits these key characteristics, from a single plant growing in soil, to the entire world ecosystem.

 The key structural components of forest ecosystems are plants, animals, microbes, soils and the atmosphere. Topography and microclimate are also important ecosystem features, but are not structural elements in the strict sense.

The key functional aspects of forest ecosystems are energy capture and biomass creation; nutrient cycling and the regulation of atmospheric and water chemistry; and important contributions to the regulation of the water cycle.

The interactions within an ecosystem involve all combinations of plant, animal and microbial interactions, interactions between organisms and the soil, and between the atmosphere and both the biotic community and the soil.

Complexity is an important attribute even though normally functioning forest ecosystems can exist at widely different levels of complexity. The importance of complexity lies in its implications for our ability to understand and predict, and therefore manage, forest ecosystems.

Forest ecosystems are continually changing. This change, initiated by external disturbance factors but largely determined by internal ecosystem processes, is vital for the maintenance of many aspects of biological diversity. In many types of forests it is essential for the long-term sustainability of the ecosystem.

"Forest stewardship" and "good, sustainable forestry" can only be defined in terms of society's desires and preferences with respect to stand and landscape-level forest conditions, functions and values. However, unless forestry is based on a respect for forest ecology and the ecological characteristics of forest ecosystems, it is very unlikely that society's long-term desires will be satisfied. Because of the long time scales of forestry, decisions about forest management must be founded on ecologically-based forecasts of ecosystem response, involving the use of ecosystem management simulation models.

Tropical Ecology

Tropical ecology is the study of all aspects of the ecology of tropical areas, which are those found approximately 23.5 degrees either side of the Equator. Notable tropical ecosystems include the rainforests of Amazonia, Africa and South East Asia, savannah grasslands and coral reefs.

Tropical ecology is a field which focuses on the ecology of the tropics, an immensely biodiverse region bounded by the Tropic of Capricorn and the Tropic of Cancer. Although people often think of the tropical rainforest when they hear the term "tropics," tropical ecosystems are actually quite varied, and include dry forests, deserts, and other types of ecosystems. Tropical ecologists, like other ecologists, are interested in the natural environment and the complex relationships of the organisms which inhabit it.

The study of tropical ecology is a very rich field within the sciences. Some types of ecosystems found in the tropics include cloud forests, dry forests, rainforests, deserts, and deciduous forests, among others, and each is very unique. The tropical climate is quite varied, ranging from very cold, dry weather on mountain peaks in the tropics to hot, humid weather in tropical valleys. Tropical ecologists can choose to study many tropical ecosystems, or to focus on a particular region or ecosystem of interest.

Ecologists look at plants, animals, insects, microorganisms, soil, and climate to learn how ecosystems form, and how they remain stable. In tropical ecology, researchers are also interested in the impact of human activities such as logging, tourism, and settlements on the environment, and ways in which humans can use the environment sustainably. Because many tropical ecosystems are very unique and irreplaceable, researchers are especially concerned about imbalances and damage in vulnerable areas.

Plant Ecology

Plant ecology is a sub-discipline of ecology focussed on the distribution and abundance of plants, and their interactions with the biotic and abiotic environment.

The Importance of Plants

The importance of plants to humans and just about all other life on Earth is staggering. Life as we know it would not be possible without plants. Why are plants so important?

- Plants supply food to nearly all terrestrial organisms, including humans. We eat either plants or other organisms that eat plants.

- Plants maintain the atmosphere. They produce oxygen and absorb carbon dioxide during photosynthesis. Oxygen is essential for cellular respiration for all aerobic organisms. It also maintains the ozone layer that helps protect Earth's life from damaging UV radiation. Removal of carbon dioxide from the atmosphere reduces the greenhouse effect and global warming.

- Plants recycle matter in biogeochemical cycles. For example, through transpiration, plants move enormous amounts of water from the soil to the atmosphere. Plants such as peas host bacteria that fix nitrogen. This makes nitrogen available to all plants, which pass it on to consumers.

- Plants provide many products for human use, such as firewood, timber, fibers, medicines, dyes, pesticides, oils, and rubber.

- Plants create habitats for many organisms. A single tree may provide food and shelter to many species of insects, worms, small mammals, birds, and reptiles.

Red-eyed tree frogs like this one live in banana trees

We obviously can't live without plants, but sometimes they cause us problems. Many plants are weeds. Weeds are plants that grow where people don't want them, such as gardens and lawns. They take up space and use resources, hindering the growth of more desirable plants. People often introduce plants to new habitats where they lack natural predators and parasites. The introduced plants may spread rapidly and drive out native plants. Many plants produce pollen, which can cause allergies. Plants may also produce toxins that harm human health.

Poison ivy causes allergic skin rashes. It's easy to recognize
the plant by its arrangement of leaves in groups of three

Why Study Plants?

Members of the plant kingdom play many crucial and sometimes surprising roles in the drama of life on Earth. You are probably familiar with some reasons plants are important. Why should you understand how plants live? Because plants play many roles, including but certainly not limited to:

1. Supplying Food and Energy

2. Maintaining Earth's Atmosphere

3. Cycling Water and Nurturing Soils

4. Contributing to Nitrogen and Other Biogeochemical Cycles

5. Interdependence with Animals

6. Interdependence with Fungi

7. Interdependence Among Plants

8. Resources for Humans

9. Aesthetics for Humans

10. Scientific Use by Humans

11. Causing Problems

More than 100,000 natural compounds come from plants, and most of these have yet to be explored. Some of the most powerful and useful compounds come from plants. Who knew they could help us unlock some of the biology's mysteries - all using an approach of mapping biological pathways.

Plant Interactions with Other Organisms

Ecology is the study of interactions of organisms with one another as well as with their environment. Plants, with their sedentary existence and need to attract pollinators or prevent herbivores from consuming them whole (because they can't run away from them), have evolved a different set of behavior patterns than have animals.

Competition

Competition results when an individual plant interferes with the needs of another plant for the same environmental resource (such as light, minerals, space) or when members of one population interfere with members of another for the same environmental resource. In plants, competition generally is indirect, through the resource, not direct, one-on-one (plants don't engage in leaf-to-leaf combat). Plants with the same life form and growth requirements are often in competition but surviving in slightly different microenvironments. This generally leads to a better utilization of the resource and, with natural selection in operation over time, a greater diversification of the community.

Secondary Metabolites

Allelopathy is a particular form of direct competition in which one plant species (or a fungus like Penicillium) produces a substance toxic to another. In some instances, the substance inhibits the development of the producer's own seeds or spores. The compounds may leach from the roots into the soil or accumulate in the ground around the plant as leaves drop and decay. Some are terpenes that volatilize and are spread through the air as aerosols. The essential oils of members of the mint family are toxic to numerous plants, as is the oil of black walnuts. Caffeine produced by tea and coffee plants inhibits the growth of seedlings of many species.

Symbiosis

Chemical warfare of another kind is waged by plants that produce secondary metabolites—chemical substances that protect the plants from being eaten by herbivores. Plants and their predators undoubtedly coevolved, with changes in one instigating reactions and further evolutionary changes in both.

Some of the metabolites are not merely deterrents, but are chemicals that imitate hormones, enzymes, or other essential compounds of animal bodies. One metabolite interferes with insect metabolism by inhibiting the juvenile growth hormone. Others, like the alkaloids morphine and cocaine, affect the human nervous system; and caffeine, although a stimulant to humans, in plants is toxic and lethal to insects and fungi. The estrogens produced by some plants have no known role in the plants, but their importance to human reproduction is well known—and a cause for concern when humans eat vegetables.

Defense substances of a different kind protect plants from bacteria and fungi attacks. These substances, called phytoalexins, act as natural antibiotics and protect the plant from bacteria and fungal pathogens when leaves are damaged or stems wounded. Nicotine in tobacco plants is synthesized in response to wounding.

In a symbiosis, two different kinds of organisms live together in an intimate and more or less permanent relationship. Lichens are the classic example of a symbiosis between a fungus and a cyanobacterium or an alga. Mycorrhizae, too, are examples of fungi and the root cells of vascular plants in a symbiosis. If the interactions between the symbionts are of mutual benefit, the symbiosis is termed a mutualism; if one partner benefits and the relationship is of no significance to the other, it is a commensalism; parasitism is a symbiosis in which one partner benefits and the other is harmed.

Mutualism. Seed plants have developed all manner of mutualisms, the most highly developed being the interactions between insects, birds, bats, and a few other animals that ensure pollination of flowers, especially by cross-fertilization. Pollinators are attracted to the flowers by colors, scents, and nectars and once on-site, all manner of structural floral adaptations insure the pollinator gets a dusting of pollen to take to the next flower it visits. The pollinator gets food, and the plant gets a messenger service more effective than chance winds.

Seed and fruit dispersal mechanisms also are well-developed, co-evolved mutualisms. Succulent edible fruits with their scents and colors are great dispersal devices geared to larger animals and often found on plants that produce seeds with hard seed coats. The coat may be so difficult for water to penetrate that germination is not possible unless some mechanical abrasion or chemical solvent is applied. The gizzard of birds is an effective grinder, and the stomach acids of mammals take off much of the seed coat before hard-coated seeds are expelled in the feces.

Parasitism. Bacteria, viruses, and fungi have not spared the plants as hosts for their parasitic lifestyle nor have vascular plants that parasitize other vascular plants. The lines among mutualism, commensalism, and parasitism are often blurred because the definitions are based on value judgments, that is, on the degrees of harm or benefit to the symbionts. About 3,000 species of vascular plant parasites are worldwide in their distribution. Some of these have lost the ability to photosynthesize entirely, but others attach to the vascular system of their hosts and divert the water and minerals in transit to their own photosynthesis.

Human Ecology

Human Ecology studies human life and human activity in different ecosystems and different cultures in the present and in the past in order to gain a better understanding of the factors which influence the interaction between humans and their environment.

The ambition to achieve a more complete view requires an integrated perspective that transcends traditional boundaries between the humanities, social sciences, natural sciences, and technology.

A fundamental issue in human ecology is how people's cultural beliefs about the nature affect and are affected by their livelihoods and the social order.

An Anthropological Perspective

While cultural beliefs come into focus in the influential modern sciences like economics, human ecologists examine the modern concepts of economic growth and technological development from an anthropological perspective. By comparing those concepts with other scientifical insights about environmental degradation, climate change and global inequality, human ecology challenges the ideas that perpetuates an unsustainable and unequal global society.

Studies in Human Ecology give you a broad and theoretically deep understanding of the interactions between man and nature in different times and in different parts of the world. Of central importance is to understand how the human relationships with the environment are influenced by their history and their place in the world system.

Human Ecology Theory

Theories of human interaction should provide a way of making sense of events that have happened in the past, and then allow us to make predictions about what may happen in the future. Human ecology theory is a way of looking at the interactions of humans with their environments and considering this relationship as a system. In this theoretical framework, biological, social, and physical aspects of the organism are considered within the context of their environments. These environments may be the natural world, reality as constructed by humans, and/or the social and cultural milieu in which the organism exists.

Human ecological theory is probably one of the earliest theories of the family and yet, it also contains many new and evolving elements that have emerged as we have begun to realize how the natural and human created environments affect our behavior, and how individuals and families in turn, influence these environments. In human ecology, the person and the environment are viewed as being interconnected in an active process of mutual influence and change.

The Origins of Human Ecological Theory

The origin of the term ecology comes from the Greek root oikos meaning "home." As a result, the field of home economics, now often called human ecology, has produced much of the contemporary research using this theoretical perspective. Margaret Bubolz and M. Suzanne Sontag (1993) attribute the concept of an ecological approach to the work of Aristotle and Plato, and then to the evolutionary theory of Darwin. They trace the word ecology to Ernest Haeckel, a German zoologist who, in 1869, proposed that the individual was a product of cooperation between the environment and organismal heredity and suggested that a science be developed to study organisms in their environment. Early home economists were major proponents of this theory as their field developed in the early twentieth century applying various disciplines to the study of the family. The theory has since been used by sociologists, anthropologists, political scientists, and economists. This work continues, with the human ecological framework being a major perspective in research and theory development in the twenty-first century.

The Family as a System

The application of systems theory is a basic tenet of human ecological theory. The family is seen as a system, with boundaries between it and other systems, such as the community and the economic system. Systems have inputs that drive various processes and actions, such as the finite amounts of money or time that families possess. They also have throughputs, which are the transformation processes that occur within the system, such as the exchange of money for the provision of an essential service, such as food, by eating in a restaurant. In addition, systems have outputs, which affect other systems, such the production of waste materials, which are byproducts of activity in the family, being returned to the larger environment. There are feedback loops from the end of the system back to the beginning, to provide both positive and negative comment back into the process and allow the system to adapt to change. In an ecosystem, the parts and the whole are interdependent.

Most theorists outline an ecosystem, most particularly a human ecosystem or a family ecosystem, as being composed of three organizing concepts: humans, their environment, and the interactions between them. The humans can be any group of individuals dependent on the environment for their subsistence. The environment includes the natural environment, which is made up of the atmosphere, climate, plants, and microorganisms that support life. Another environment is that built by humans, which includes roads, machines, shelter, and material goods. As Sontag and Bubolz (1996) discuss, embedded in the natural and human-built environments is the social-cultural environment, which includes other human beings; cultural constructs such as language, law, and values; and social and economic institutions such as our market economy and regulatory systems. The ecosystem interacts at the boundaries of these systems as they interface, but also can occur within any part of an ecosystem that causes a change in or acts upon any other part of the system. Change in any part of the system affects the system as a whole and its other subparts, creating the need for adaptation of the entire system, rather than minor attention to only one aspect of it.

There are also systems nested within systems, which delineate factors farther and farther from individual control, and that demonstrate the effects of an action occurring in one system affecting several others. Urie Bronfenbrenner's analysis of the systems such as the microsystem, mesosystem, exosystem, and macrosystem are an integral part of the theory. The microsystem is our most immediate context, and for most children, is represented by their family and their home. Young children usually interact with only one person until they develop and their world expands. The mesosystem is where a child experiences reality, such as at a school or childcare setting. Links between the institutions in the mesosystem and the child's family enhance the development of academic competence. The exosystem is one in which the child does not participate directly, but that affects the child's experiences. This may be a parent's workplace and the activities therein, or bureaucracies that affect children, such as decisions made by school boards about extracurricular activities. Our broadest cultural identities make up the macrosystem. This system includes our ideologies, our shared assumptions of what is right, and the general organization of the world. Children are affected by war, by religious activities, by racism and sexist values, and by the very culture in which they grow up. A child who is able to understand and deal with the ever-widening systems in his or her reality is the product of a healthy microsystem.

Bubolz and Sontag (1993) outline five broad questions that are best answered using this theory, which is helpful in deciding areas where the theory can make a useful contribution to our knowledge. These are:

1. To understand the processes by which families function and adapt—how do they ensure survival, improve their quality of life, and sustain their natural resources?

2. To determine in what ways families allocate and manage resources to meet needs and goals of individuals and families as a group. How do these decisions affect the quality of life and the quality of the environment? How are family decisions influenced by other systems?

3. How do various kinds and levels of environments and changes to them affect human development? How does the family system adapt when one or more of its members make transitions into other environmental settings, such as day care, schools, and nursing homes?

4. What can be done to create, manage, or enhance environments to improve both the quality of life for humans, and to conserve the environment and resources necessary for life?

5. What changes are necessary to improve humans' lives? How can families and family professionals contribute to the process of change?

A basic premise of a human ecological theory is that of the interdependence of all peoples of the world with the resources of the earth. The world's ecological health depends on decisions and actions taken not only by nations, but also by individuals and families, a fact that is increasingly being realized. Although the concept of a family ecosystem is not a precise one, and some of the terms have not been clearly and consistently defined, a human ecological theoretical perspective provides a way to consider complex, multilevel relationships and integrate many kinds of data into an analysis. As new ways of analyzing and combining data from both qualitative and quantitative dimensions of interconnected variables develop, this theoretical perspective will become more precise and continue to enhance understanding of the realities of family life.

Insect Ecology

Insect ecology is a field which focuses on the study of the interaction between insects and the environment. While laypeople may think of insects primarily in the form of irritating bugs like mosquitoes which ruin an evening barbecue, insects are actually very important to the natural environment, and they play a number of roles in the environment, from angel to villain. Several disciplines are brought together in insect ecology, including entomology, ecology, and microbiology.

Insects are a critical part of the circle of life in the environment. When animals and plants die, several important insect species start the process of breaking down the organic material so that it can be digested by even smaller bacteria and fungi. Insects also act as pollinators, ensuring the survival of plant species, and they can play a more menacing role as vectors for disease. Insects can even help in criminal investigations.

Insect ecology can include the study of insect behavior, the impact of human activities on insect populations and the ecosystem at large, the role of insects in human history, and what happens when insects are absent from an environment. Insect ecologists are also interested in issues like controlling dangerous insects, identifying and studying insects which carry disease, and the impact of introduced non-native insect species on the environment.

Community Ecology

Community ecology is the process by which a group of organisms which live in the same location interact. There is direct interaction, which takes the form of symbiosis, competition and predation, which are the most easily notable. There is also indirect interaction, such as reproduction, foraging patterns and decaying. Every organism is at its most basic state could be a consumer in some situations, and a producer in others. The culmination of all these interactions is what defines a community and what differentiates one from another. Insects often play several roles in these communities, though these roles vary widely based on what species is present.

Decomposers

Decomposer insects are ones that feed on dead or rotten bodies of plant or animal life. These insects are called saprophages and fall into three main categories; those that feed on dead or dying plant matter, those that feed on dead animals (carrion), and those that feed on excrement (feces) of other animals. As dead plants are eaten away, more surface area is exposed, allowing the plants to decay faster due to an increase in microorganisms eating the plant. These insects are largely responsible for helping to create a layer of humus on the soil that provides an ideal environment for various fungi, microorganisms and bacteria. These organisms produce much of the nitrogen, carbon, and minerals that plants need for growth. Carrion feeders include several beetles, ants, mites, wasps, fly larvae (maggots), and others. These insects occupy the dead body for a short period of time but rapidly consume and/or bury the carcass. Typically, some species of fly are the first to eat the body, but the order of insects that follows is predictable and known as the faunal procession. Many dung beetles and manure flies are attracted to the smell of animal feces. The adults often lay eggs on fresh excrement and the larvae will feed on the organic matter. Many species of dung-feeders have evolved so they will only feed on feces from a specific species. There is even a type of dung-beetle that will roll feces into a ball, push it into a pre-dug hole, laying an egg in that dung and then cover it with fresh dirt to provide a perfect nursery for their larvae.

Carnivores

Carnivorous insects survive by eating other living animals, be it through hunting, sucking blood, or as an internal parasite. These insects fall into three basic categories: predators, parasites, and parasitoids. Predatory insects are typically larger as their survival is dependent upon their ability to hunt, kill/immobilize, and eat their prey. There are several exceptions to this though, with ants being the most notable. Ants, and other colony insects, can use their sheer numbers to overwhelm their prey even if the ants are significantly smaller. They often have specialized mandibles (mouth parts) for this task, some causing excruciating pain, paralysis, or simply having a high bite force. The insects who live on their own though, must be able to reliably bring down their prey and as such have developed a myriad of unique hunting methods. Some actively travel, seeking out their prey while others wait in an ambush. Others may release chemicals to attract specific creatures and others still will eat anything they can. Parasites infest the victim's body and eat it from the inside out. The presence of the parasite is often not noticed by the host as the size discrepancy is typically so vast. Parasites vary widely in how they survive in their host, some complete their full life cycle within the body while others may only stay in for the duration of their larval stage. There is as great of variation in methodology and species in parasites as in any other type of insect even if it may not seem so at first. The most threatening parasites to humans though, are ones that live outside the host and consume the hosts blood. These species transmit virus, disease, and even other, smaller parasites to the host, spreading these throughout the populations of many third world countries with poor health care. A subcategory of parasites, called parasitoids, is one that feeds on the host body so much so that the host is eventually eaten. One species of wasp, the spider wasp, will paralyze spiders before bringing them back to their nest and injecting it with a wasp larvae. The larvae will eat its way out, secreting a numbing and paralyzing agent until there is nothing left of the spider other than the exoskeleton then go through a metamorphism and become an adult wasp.

Herbivores

Herbivores are insects that feed on living plant matter or the products of a plant. These insects may eat essential parts of the plant, such as the leaves or sap, or they may survive on the pollen and nectar produced by the plant. Herbivorous insects often use olfactory or visual cues to determine a potential host plant. A visual cue could simply be the outline of a certain type of leaf, or the high contrast between the petals of a flower and the leaves surrounding it. These are typically associated with the olfactory signal an insect may receive from their intended meal. The olfactory que could be the scent of the nectar produced by a flower, a certain chemical excreted to repel unwanted predators, or the exposed sap of a cherry tree. either of these two senses could be the driving force behind an insect choosing to consume a certain plant, but it is only after it takes the first bite, and the confirmation of this food is made by its sense of taste, that it truly feeds. After a herbivorous insect is finished feeding on a plant, it will either wait there until hungry again, or move on to another task, be it finding more food, a mate, or shelter. Herbivorous insects bring significantly more danger to a plant than simply that of consumption, they are among the most prominent disease carrying creatures in the insect world. There are numerous diseases, fungi, and parasites that can be carried by nearly any herbivorous insect many of which fatal to the plant infected. Some diseases even produce a sweet smelling, sticky secretion from the infected plant to attract more insects and spread farther.

Landscape Ecology

Landscape ecology falls under the broader heading of ecology, focusing on the correlations between spatial patterns and landscape characteristics. Landscape ecology also considers the effects of land use on natural resources such as forests, wetlands, grasslands, lakes, rivers, streams, and other natural settings. These ecological considerations make it possible to manage landscape use to minimize negative effects on the environment.

A landscape is an area of land of any size that contains a specific pattern that impacts and is impacted by various ecological processes. Landscape ecology studies these specific patterns and how various elements within the landscape interact to cause change. Landscape ecology seeks to use the theories from these observations to solve environmental challenges. Landscape ecology involves three standard characteristics: landscape structure, landscape function, and landscape change.

The structure of a landscape involves the spatial arrangement of various elements present in the landscape. The function of a landscape involves movement, both of animals living in the landscape and of landscape elements and ecosystems in an area, such as water, plants, wind, and materials. The change of a landscape involves observation of how spatial arrangement or the function of a landscape changes over time.

Landscape ecology involves the study of patterns in landscape structures. Patches are habitat areas that vary in size, shape, number, and location. Patches can cluster in an area with a number of them coexisting in small proximity, or they may occur in more isolated fashion. Corridors separate patches, serving as boundaries. Corridors can have a variety of characteristics, such as straight or curvy perimeters or a narrow or wide width. Corridors may also be continuous or disconnected. The matrix is the landscape that surrounds patches and corridors. Matrix areas could be farmland, for example.

Landscape Ecology Principles

The science of landscape ecology actually includes numerous sciences under the broad umbrella of landscape ecology. These sciences include antropology, geobotany, geomorphology, soil science, and zoology. In combining these areas of science to make up landscape ecology, it is possible to observe how interactions between these separate areas affect various ecological processes.

One of the main principles of landscape ecology notes the change that occurs in spatial arrangements, either as a result of human actions or from natural processes. Scientists study the relationship between human activity and landscape pattern, noting the process by which change occurs.

Human activity and flawed landscape designs have led to a variety of environmental issues that threaten the earth, including air and water pollution, the spread of invasive species, loss of biodiversity, and significant climate changes. Urban development, industrialization, and disappearing ecoscapes are principle threats to the natural environment. Natural ecosystems such as tropical rainforests, deciduous forests, grasslands, and wetlands have been disrupted, with alarming results. Animal and insect species have declined, native plants are disappearing, and greenhouse gas emissions have caused climatic changes. Applying the principles of landscape ecology enables management of land use, animal and plant populations, and natural disturbances.

Landscape design should harmoniously connect human innovation with the nature that surrounds it. Design that considers responsible use of natural resources, environmental sustainability, and minimal stress to natural ecosystems will benefit everyone. Not only can human economy flourish, but the environment's natural ecosystems can be maintained and even enhanced.

Spatial Heterogeneity

Landscape ecology might be defined best by its focus on spatial heterogeneity and pattern: how to characterize it, where it comes from, how it changes through time, why it matters, and how humans manage it. As such, landscape ecology has five central themes:

Landscape Ecology.....*focus on spatial heterogeneity and pattern*

- How to characterize it...
- Where it comes from...
- How it changes over time...
- Why it matters...
- How humans mange it...

- Detecting pattern and the scale at which it is expressed, and summarizing it quantitatively.

- Identifying and describing the agents of pattern formation, which include the physical abiotic template, demographic responses to this template, and disturbance regimes overlaid on these.

- Characterizing the changes in pattern and process over space and time; that is, the dynamics of the landscape, and summarizing it quantitatively. An interest in landscape dynamics necessarily invokes models of some sort--because landscape are large and they change (usually!) over time scales that are difficult to embrace empirically.

- Understanding the ecological implications of pattern; that is, why it matters to populations, communities, and ecosystems – and this is the stuff of conservation biology and ecosystem management.

- Managing landscapes to achieve human objectives.

Broad Spatial Extents

Landscape Ecology.....*focus on broader spatial extents*

- Landscape ecology OFTEN focuses on spatial extents that are much larger than those traditionally studied in ecology...but the emphasis is on spatial pattern at the relevant scale.

Landscape ecology is distinguished by its focus on broader spatial extents than those traditionally studied in ecology. This stems from the anthropocentric origins of the discipline. Initial impetus for the discipline came from the geographers aerial view of the environment, for example, the patterns in the environment visible from an aerial photograph. The focus on large geographic areas is consistent with how humans typically see the world—through a coarse lens. However, modern landscape ecology does not define, a priori, specific scales that may be universally applied; rather, the emphasis is to identify scales that best characterize relationships between spatial heterogeneity and the process of interest.

The Role of Humans

Landscape ecology is often defined by it focus on the role of humans in creating and affecting landscape patterns and process. Indeed, landscape ecology is sometimes considered to be an interdisciplinary science dealing with the interrelation between human society and its living environment. Hence, a great deal of landscape ecology deals with 'built' environments, where humans are the dominant force of landscape change. However, modern landscape ecology, with its emphasis on the interplay between spatial heterogeneity and ecological process, considers humans as one of many important agents affecting landscapes, and emphasizes both natural, semi-natural, and built landscapes.

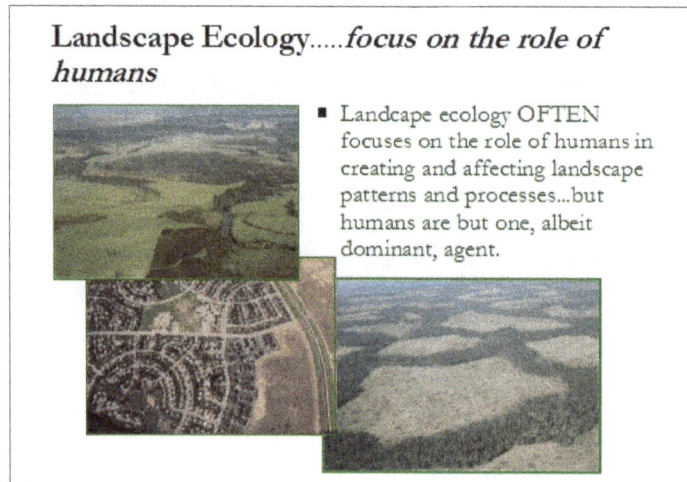

Landscape Ecology.....*focus on the role of humans*

- Landcape ecology OFTEN focuses on the role of humans in creating and affecting landscape patterns and processes...but humans are but one, albeit dominant, agent.

Why is Landscape Ecology Important to Resource Managers?

Landscape ecology (or a landscape perspective) with its focus on spatial patterns is important to resource managers because: 1) ecosystem context matters, 2) ecosystem function depends on the interplay of pattern and process, and 3) because human activities can dramatically alter landscape context and the relationship between patterns and processes, resource managers have a stewardship responsibility to understand and manage these impacts – more pragmatically, resource managers have a policy and legal mandate to include a landscape perspective in resource management decisions.

Because Ecosystem Context Matters

Landscape ecology is founded on the principle that ecosystem composition, structure and function partially depend on the spatial (and temporal) context of the ecosystem (i.e., its landscape context); i.e., that what we observe ecologically at any particular location is affected by what is around that location. This shift in perspective from the site to the site embedded in a landscape context has profound implications for resource management. Let's consider a couple of examples:

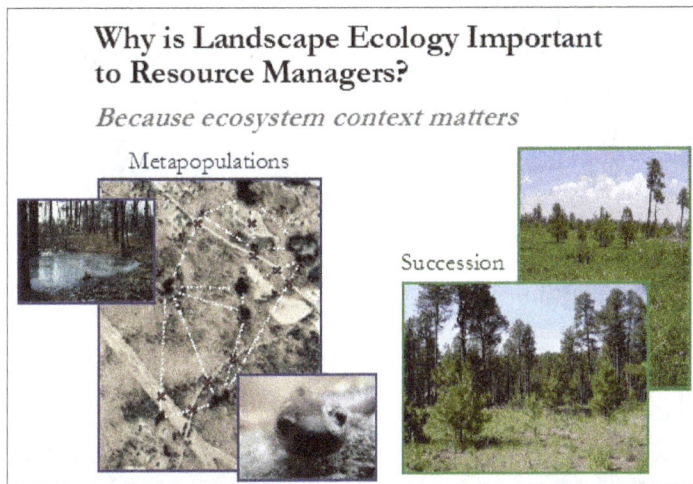

Why is Landscape Ecology Important to Resource Managers?

Because ecosystem context matters

Metapopulations

Succession

- Metapopulations.–Metapopulation s depend on the number and spatial arrangement of habitat patches – where the probability of a habitat patch being occupied at any time is at least partially dependent on its proximity to other habitat patches. Focusing management on the individual site, in this case, without consideration of its landscape context, can have disastrous consequences for the population.

- Forest succession.–Neighborhood effects can play an important role in determining the successional response following a disturbance. For example, edge effects that modify the distribution of energy and water and the plant species composition of the immediate neighborhood (which can influence the relative abundance of propagules) can exert a strong influence on succession in forest gaps and in larger openings, e.g., via wave-form succession. Ignoring these effects can lead to undesirable outcomes, including an unwanted shift in species composition or an inadequate recovery of vegetation altogether.

Because ecosystem function depends on the interplay of pattern and process

Landscape ecology is also founded on the principle that spatial patterns affect ecological processes, which in turn affect spatial patterns. This interplay of spatial pattern and process is in fact the overarching focus of landscape ecology. While it can be argued that "ecology" has always sought to explain the relationship between pattern and process, it is safe to assert that "landscape" ecology has shifted the focus to pattern-process relationships over broad spatial extents and emphasized the role of humans in creating and affecting these relationships. This shift has profound implications for resource managers. Let's consider a couple of examples:

- Habitat fragmentation. –Disruption of habitat connectivity is a major impact of human activities on plant and animal populations and one of the leading causes of the biodiversity crisis. Anthropogenic landscape elements (e.g., roads, developed land, dams) can function as impediments to the movement of organisms across the landscape, and the cumulative impacts of these impediments over broad spatial extents can be devastating.

- Alteration of disturbance regimes.–Disruption of natural disturbance regimes has longlasting ecological and socioeconomic impacts. For example, disruption of fuel mass and continuity by human land use practices (e.g., livestock grazing) over broad spatial extents can dramatically alter fire regimes in fire-dependent ecosystems, leading to shifts species distributions and community structure and serious socio-economic consequences (e.g., catastrophic fires resulting in loss of life and property).

Because there is a policy and legal mandate to include a landscape perspective

Lastly, and more pragmatically, there is a policy and legal mandate to include a landscape perspective into resource management decisions. All federal land management agencies have formally adopted "ecosystem management" as the overarching resource management paradigm, and a landscape perspective (and all that it subsumes) is one of the pillars of the ecosystem management approach. More specifically, and more tangibly, a landscape perspective plays a critical role in the 2005 Forest Service Planning Rule (36 CFR Part 219) and the subsequent regulations to implement that rule.

Applied Ecology

Applied ecology is a subfield within ecology, which considers the application of the science of ecology to real-world (usually management) questions. It is an integrated treatment of the ecological, social, and biotechnological aspects of natural resource conservation and management.

It is also called ecological or environmental technology. Applied ecology typically focuses on geomorphology, soils, and plant communities as the underpinnings for vegetation and wildlife (both game and non-game) management.

Aspects of applied ecology include:

Yosemite National Park in the United States

- Agro-ecosystem management
- Biodiversity conservation
- Biotechnology
- Conservation biology
- Ecosystem restoration
- Habitat management
- Invasive species management
- Protected areas management
- Rangeland management
- Restoration ecology

Major journals in the field include:

- Journal of Applied Ecology
- Ecological Applications
- Applied Ecology and Environmental Research

Related organizations include:

- Ecological Society of America (The Americas)
- Society for Ecological Restoration (Global)

- Institute for Applied Ecology (USA)

- Kazakh Agency of Applied Ecology

- Öko-Institut (Institute for Applied Ecology) (in Germany)

Biogeography

Biogeography refers to the distribution of various species and ecosystems geographically and throughout geological time and space. Biogeography is often studied in the context of ecological and historical factors which have shaped the geographical distribution of organisms over time. Specifically, species vary geographically based on latitude, habitat, segregation (e.g., islands), and elevation. The subdisciplines of biogeography include zoogeography and phytogeography, which involve the distribution of animals and plants, respectively.

Types of Biogeography

There are three main fields of biogeography: 1) historical, 2) ecological, and 3) conservation biogeography. Each addresses the distribution of species from a different perspective. Historical biogeography primarily involves animal distributions from an evolutionary perspective. Studies of historical biogeography involve the investigation of phylogenic distributions over time. Ecological biogeography refers to the study of the contributing factors for the global distribution of plantand animal species. Some examples of ecological factors that are commonly studied include climate, habitat, and primary productivity (the rate at which the plants in a particular ecosystem produce the net chemical energy). Moreover, ecological biogeography differs from historical biogeography in that it involves the short-term distribution of various organisms, rather than the long-term changes over evolutionary periods. Conservation biogeography seeks to effectively manage the current level of biodiversity throughout the world by providing policymakers with data and potential concerns regarding conservation biology.

Biogeography Support Evolution

Biogeography provides evidence of evolution through the comparison of similar species with minor differences that originated due to adaptations to their respective environments. Over time, the Earth's continents have separated, drifted apart, and collided, resulting in the creation of novel climates and habitats. As species adapted to these conditions, members of the same species that had been separated geographically diverge, resulting in the eventual formation of distinct species. This knowledge is important, as by understanding how adaptations occurred in response to changing environments in the past, we can apply this knowledge to the future.

One of the most famous examples of biodiversity in support of evolution is Charles Darwin's study of finches on the Galapagos Islands, which resulted in his book On the Origin of Species. Darwin noted that the finches on the mainland of South America were similar to those located on the Galapagos Islands; however, the shape of the bills differed depending on the type of food available on each island. The islands had once been a part of the South American mainland, but the two land masses were subsequently separated and drifted apart. The result was the creation of novel habitats and food sources available for the species residing in each of these regions. Therefore, each finch species had adapted to the local environment through the selection of alleles which promoted survival, eventually resulting in speciation. Islands are excellent for the study of biogeography because they consist of small ecosystems that can easily be compared to those of the mainland and other nearby regions. Moreover, since they are an isolated region, invasive species and the associated consequences for other organisms within the ecosystem can be readily studied. By studying such changes over time, the evolution of distinct species and ecosystems becomes apparent.

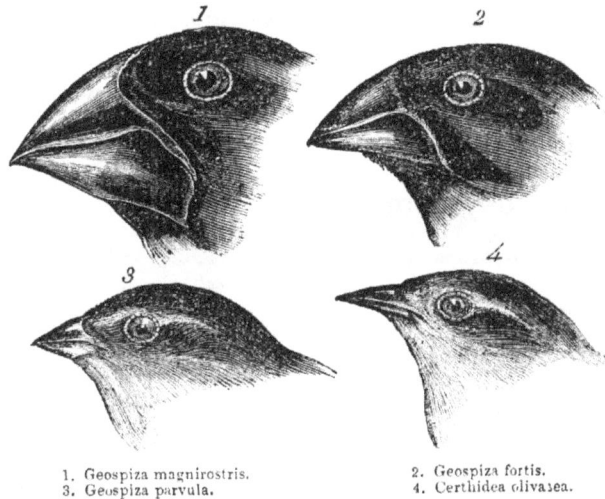

1. Geospiza magnirostris.
3. Geospiza parvula.
2. Geospiza fortis.
4. Certhidea olivasea.

Darwin's finches

Urban Ecology

The emerging science of urban ecology, a subdiscipline of ecology that examines the interactions between organisms and the human-dominated ecosystems in which they reside, may provide additional solutions to urban environmental problems. Ecologists first began to perform comprehensive studies of plants, animals, soils, and environmental conditions in cities shortly after World War II, when there were many vacant sites within European cities. They began a tradition of examining open spaces, which supported volunteer plant communities and the animal populations associated with them. Ecological planning emerged as a professional discipline that applied knowledge of the open spaces in urban areas in an effort to enhance biological diversity and amenities originating from green patches in cities.

A different tradition originated in sociology, which applied ecological concepts such as competitionand succession to the human communities of cities. In the 1990s, ecological research in urban areas burgeoned. To deal with the complex mosaic of land uses that now make up cities, suburbs, and exurbs, the traditions were combined, along with knowledge taken from other disciplines, to establish a comprehensive ecological approach to the study of urban ecosystems.

Urban ecology has grown increasingly important as a result of the migration of most of the global human population to cities. One of the by-products of this unprecedented phenomenon is that the world's urban areas are expanding into environmentally sensitive locations, where they alter ecosystem structure through pollution and land-use conversion of natural habitats. The knowledge gained from studying biological communities in cities may assist in the development of improved urban design and decision making in dealing with such problems.

Urban Ecosystem

Urban ecosystem, any ecological system located within a city or other densely settled area or, in a broader sense, the greater ecological system that makes up an entire metropolitan area. The largest urban ecosystems are currently concentrated in Europe, India, Japan, eastern China, South America, and the United States, primarily on coasts with harbours, along rivers, and at intersections of transportation routes. Large urban areas have been features of the industrialized countries of Europe and North America since the 19th century. Today, however, the greatest urban growth occurs in Africa, South and East Asia, and Latin America, and the majority of megacities (that is, those with more than 10 million inhabitants) will be found there by 2030.

The Structure of Urban Ecosystems

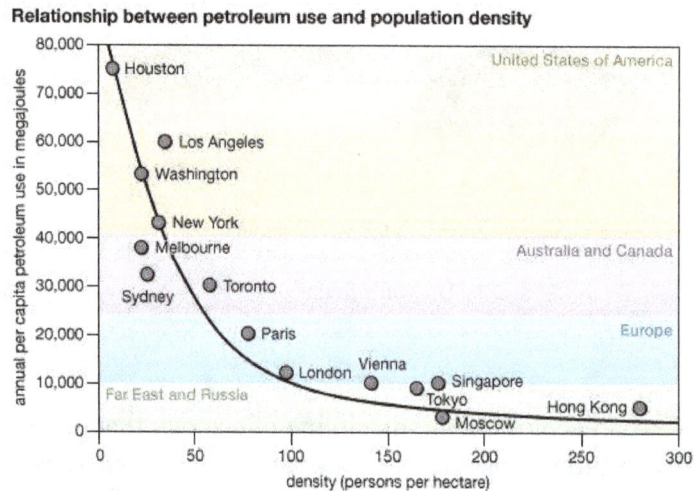

petroleum use and population density graph showing the relationship between per capita petroleum use and urban population density for selected cities

Urban ecosystems, like all ecosystems, are composed of biological components (plants, animals, and other forms of life) and physical components (soil, water, air, climate, and topography). In all ecosystems these components interact with one another within a specified area. In the case of urban ecosystems, however, the biological complex also includes human populations, their demographic characteristics, their institutional structures, and the social and economic tools they employ. The physical complex includes buildings, transportation networks, modified surfaces (e.g., parking lots, roofs, and landscaping), and the environmental alterations resulting from human decision making. The physical components of urban ecosystems also include energy use and the import, transformation, and export of materials. Such energy and material transformations involve not only beneficial products (such as transportation and housing) but also pollution, wastes, and excess heat. Urban ecosystems are often warmer than other ecosystems that surround them, have less infiltration of rainwater into the local soil, and show higher rates and amounts of surface runoff after rain and storms. Heavy metals, calcium dust, particulates, and human-made organic compounds (e.g., fertilizers, pesticides, and contaminants from pharmaceutical and personal care products) are also concentrated in cities.

The expansion of large urban areas results in the conversion of forests, wetlands, deserts, and other adjacent biomes into areas devoted to residential, industrial, commercial, and transportational uses. Such conversion may result in the production of barren land. In addition, the conversion process fragments remaining wild or rural ecosystems into ever-smaller patches, and relatively high amounts of suboptimal habitat are found at the boundaries between the remaining native ecosystems and those that have been modified for human use. Such "edge habitats" inhibit specialist plant and animal species—that is, species that can tolerate a narrow range of environmental conditions. In addition, nonurban ecosystems downwind and downstream of urban ecosystems are subjected to high loads of water pollution, air pollution, and introduced exotic species.

Aerial view of a residential subdivision in Quebec

Urban animal communities tend to be dominated by medium-size generalists, such as raccoons, coyotes, opossums, skunks, foxes, and other animals capable of surviving across a wide range of environmental conditions. In contrast, nonurban ecosystems tend to contain specialist species and animals that vary across a broader range of sizes. Urban habitats tend to be dominated by introduced plant and animal species that have a long history of association with humans and that show adaptations to urban conditions. For example, birdsong in urban areas often has a higher pitch and louder volume than is heard in nonurban populations of the same species. Louder, higher-pitched song allows birds to communicate in spite of the greater noise levels found in and around cities and suburban transportation corridors.

Compared with plant and animal communities found in wild and rural ecosystems around the world, biological communities found in different urban areas tend to be similar to one another. This ecological similarity is a by-product of the structural similarities among urban environments(comparable building types, landscape designs, and infrastructure) and of the intentional or accidental introduction of similar species into cities, suburbs, and exurban areas and the water and nutrient subsidies provided by people and their activities. Introduced groups include rodents, earthworms, shade trees, weeds, and insect pests. In addition, animal populations in urban areas sometimes show evidence of genetic differentiation from rural populations of the same species.

Abiotic Component

In the environment, there are external factors that really affect organism living on it. And one of these factors is Abiotic factors or the nonliving variables such as wind, ocean, day length, rainfall, temperature and ocean current. Abiotic factors influence the flow of interaction in an environment so it is an important move to study their effects on living organisms.

Abiotic or Nonliving things have a vital role in maintaining the balance of the ecosystem. The factors of Abiotic have varied components and aspect in the physical environment on how they affect biotic factors. Below are some of the observations that will help you to learn further about Abiotic factors.

1. Bamboo can stand on strong winds while banana plant cannot for it don't have hard trunk and doesn't sway with the blowing wind.

2. Cogon thrives well in abundant sunlight while ferns are much on shades that is why they are shade-loving plants.

3. Coconut grows well in warm climates while Pine trees in cold climates.

4. Cacti can withstand arid places like deserts while mosses can't for they are moisture-loving plants.

The descriptions and examples above are some of the effects of climate on the growth and thriving abilities of plants, more specifically light, temperature, moisture and wind. The soil is another aspect of the physical environment that we should also consider for the characteristics of a soil determines what type of organism/living things can live. Below are some of the things that must be considered.

1. The nutrients in the soil

2. Acidity level of the soil

3. Moisture content of the soil

Abiotic like stones

The amount of water that the soil can hold and the amount of minerals that can drain away is affected by the acidity of soil and the size of particles on it. Topography is also one of the aspects of the physical environment. Below are some of the observations that can make these things clear out, in the aspect of topography, and the effects on the distribution and growth of an organism/living things.

1. Most of the mossy forests are found at elevation above 1520 meters and not in the lowlands. Cloud rats are found in high mountains unlike the lowland field rats.

2. Plants which need large amount of water are found in lowlands or along gentle slopes while plants which can tolerate little moisture grow along steep slopes.

3. It was also observed that mountains slopes which are oriented facing the sun generally have thicker plant growth than those on the shaded side.

These are some observations that illustrate the effect of topography of the land on plant and animal life, more specifically altitude or elevation angle of slope and orientation of the slope.

Climate (Factors)	Soil (Factors)	Topography (Factors)
Light	Nutrients in the soil	Altitude or elevation
Temperature	Acidity of the soil	Angle of the slope
Moisture	Moisture content	Orientation of the slope
Air/wind	Of the soil	

Four Main Abiotic Factors

Abiotic or non-living things contribute to the physical components of the environment such as water, soil, air, heat and light. They are continually subjected in different situations in the physical environment such as erosion, typhoons, volcanic events, ocean current and etc. It can cause extinction any type of organism as a threat and may cause alteration to form new species and hybrid as well.

1. Water - it is an essential part in the environment where organism can find their food, shelter, a way to escape from predator, and a place for marine life. Bodies of water act as "heat sink" to slow down the large temperature changes creating a more stable environment. The bodies of living things are almost made up of water, likewise, the earth is 70% water and 30% land. Helps in the process of photosynthesis and other cycles in nature.

2. Air - is a mixture of several gases; it is the second key abiotic factors that contribute to ecosystem, where birds can fly and seeds can be disperse. Air composes of different gases: 78% nitrogen, 21% oxygen and the remaining 1%contains mostly of hydrogen, carbon diode, water, helium, argon and krypton. They endlessly circulate where all life depends.

3. Soil - is the third major abiotic key for physical environment. Basic medium for land base ecosystem where plants grow in and some organism lives on it. Soil is a natural reservoir for the inorganic mineral elements such as iron, zinc, calcium and phosphorus. It also contains humus which is certainly come from human and decayed plants. Soil which contains more humus is support more plant and animal life. It affects countless organisms.

4. Heat - is the fourth abiotic major factors that affects physical environment. Most heat energy here on earth surface originates and come from the sun. The effect of heat are very obvious, tropical environment support different organism than cold environment. Temperature affects what kind of organism can live in a certain place.

Biotic Component

Biotic Factors of Ecosystem: Producers, Consumers and Decomposers

Ecosystem is composed of biotic factors of a community of living organism interacting with one another which we can see in food chains/webs. These diverse organisms stay together because of the need of food. Population is referred to as a collection of same species. This population plays a role in maintaining the equilibrium in the ecosystem. Populations for food-- the roles played by the population in this feeding relationship are classified into the three major biotic components of an ecosystem namely:

- Producers

- Macro-consumer

- Decomposer

What are Producers?

These are organisms called producers can manufacture their own food from simple organic substances through the process of "photosynthesis". They are often said to be "autotrophs" which comes from the Greek word "autos" which means self and "trophikos" which means nursing that refers to nutrition.

The name "autotrophic" pertains to "self-nourishment or self-feeding." Producers manufacture foods by converting sunlight to energy in the process of photosynthesis. One good example is the chlorophyll-bearing plants and photosynthetic bacteria. But some producers manufacture food without the aid of sunlight by just using chemical energy of simple inorganic substances like mushrooms and the process is called "chemosynthesis" and the organism is known as chemosynthesis bacteria. Plants photosynthetic and chemosynthetic bacteria are autotrophic organisms that produce food in the ecosystem.

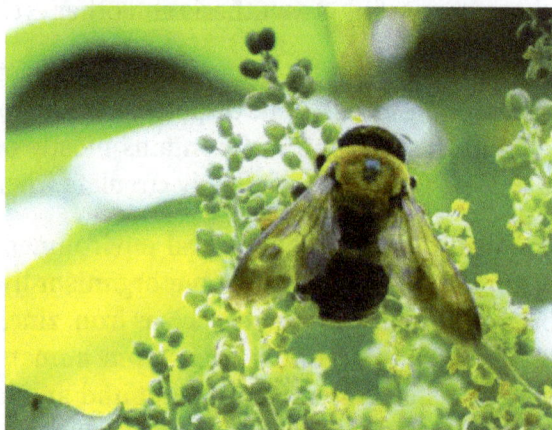

Producers are mainly of the first stage of food chain for they are the one of sustain and produce food for other living creatures or biotic factors.

Macroconsumers

This is mostly composed of herbivores, carnivores and omnivores. Macroconsumers are large consumers like animals depends food on other organism. Hence, macroconsumers like decomposers may also described as heterotrophic. The hetero which means other because the term describe organisms the feed on others, macroconsumers are grouped into three according to their food preferences;

1. Plants-eaters (herbivores) – These are consumers who eats plant leaves, flowers, stems, roots and etc. Some of the few examples of herbivores are carabao, horses and goats. Plant-eaters are vegetarian animals so meaning they do not eat meat of other animals.

2. Flesh-eater (carnivores) – These are animals that eat meat of other animals and some examples of carnivorous animals are dog, snake and hawk. They don't eat plants but only eat meat of the animals. It comes from the Greek word "Carni" which means meat and the other means eaters.

3. Variety-eaters (Omnivores) –These are consumers who actually eat either plants or animal meats. Some examples of omnivorous animals are rat, chicken and man that eat both plants and animals.

Saprophytes

Decomposers

Decomposers refer to small consumers like bacteria, fungi and worms that cause the decay of dead organism. They are sometimes called to as microconsumer because many of them are microorganisms that are too small to be seen by the naked eye.

They are also described as heterotrophic because the feed on other organism. Some decomposers like fungi are saphrotrophic. This means fungi take in food by absorbing dissolved organic substances that are products of decay.

Decomposers break down the complex substances in bodies of dead plants and animals into simpler materials. They are the final consumers of a biotic community. They return to the nonliving environment the materials which were originally absorbed by plants from the soil.

Ecological Trap

An ecological trap occurs when an organism prefers or chooses one habitat over another but the chosen habitat is lower in habitat quality. Often there is a cue that the organism misperceives as a sign of good habitat. The organism is sacrificing a more suitable habitat for one that is detrimental to their fitness. Ecological traps are often associated with anthropogenic changes to the environment. Humans may change the environment to present different or exaggerated cues to organisms.

There are few well demonstrated examples of ecological traps. It is difficult to show that one is occurring because you need to provide evidence firstly, that one habitat is being chosen over another and secondly, that that habitat is causing survival and/or reproduction to be lower than in the other.

One example is of dragonflies that require a water source in order to lay their eggs. They find the water source by sensing polarised light. However, black tarmac on roads send out stronger polarised light than natural water sources and therefore attract dragonflies through that cue. Dragonflies lay their eggs on the roads and consequently suffer high mortality and low reproductive success. Another, amusing example used by an author is of little moths being attracted to his pot of beer and then drowning in it.

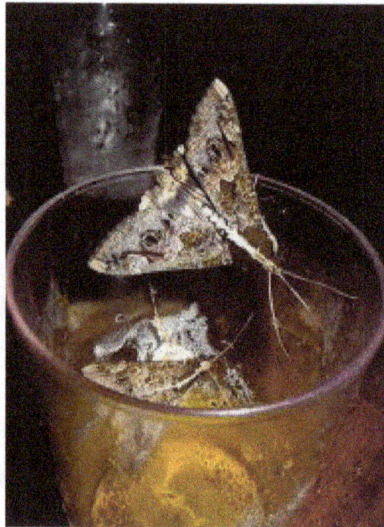

Unfortunately, not all examples are so straight forward. Direct mortality is not so easy to detect in many natural environments. What makes it harder is that the preferred habitat continues to be replenished from source populations because organisms are more drawn there. So population doesn't decline

Ecosystem

An ecosystem can be categorized into its abiotic constituents, including minerals, climate, soil, water, sunlight, and all other nonliving elements, and its biotic constituents, consisting of all its living members. Linking these constituents together are two major forces: the flow of energy through the ecosystem, and the cycling of nutrients within the ecosystem.

The fundamental source of energy in almost all ecosystems is radiant energy from the Sun. The energy of sunlight is used by the ecosystem's autotrophic, or self-sustaining, organisms. Consisting largely of green vegetation, these organisms are capable of photosynthesis—i.e., they can use the energy of sunlight to convert carbon dioxide and water into simple, energy-rich carbohydrates. The autotrophs use the energy stored within the simple carbohydrates to produce the more complex organic compounds, such as proteins, lipids, and starches, that maintain the organisms' life processes. The autotrophic segment of the ecosystem is commonly referred to as the producer level.

Organic matter generated by autotrophs directly or indirectly sustains heterotrophic organisms. Heterotrophs are the consumers of the ecosystem; they cannot make their own food. They use, rearrange, and ultimately decompose the complex organic materials built up by the autotrophs. All animals and fungi are heterotrophs, as are most bacteria and many other microorganisms.

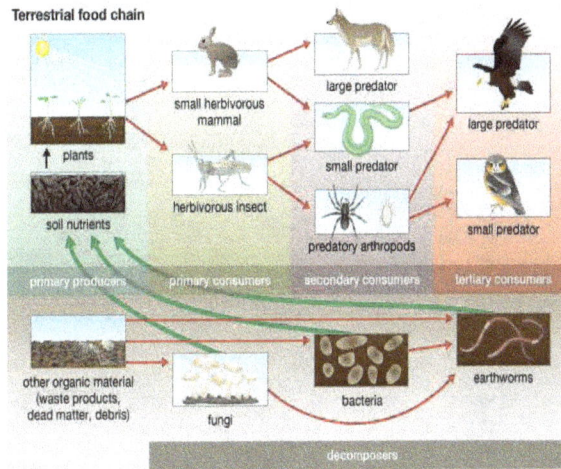

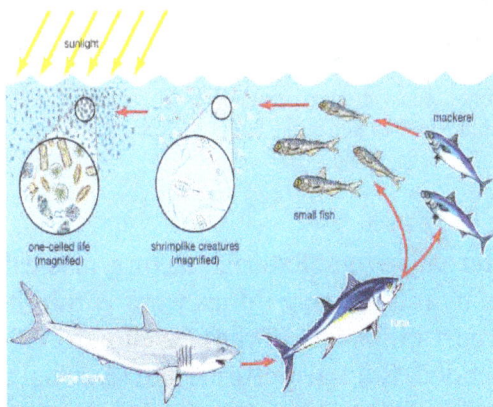

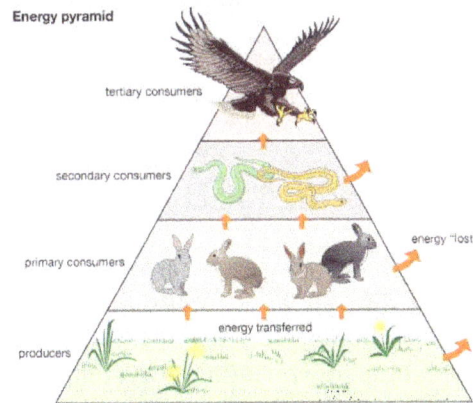

Together, the autotrophs and heterotrophs form various trophic (feeding) levels in the ecosystem: the producer level, composed of those organisms that make their own food; the primary consumer level, composed of those organisms that feed on producers; the secondary consumer level, composed of those organisms that feed on primary consumers; and so on. The movement of organic matter and energy from the producer level through various consumer levels makes up a food chain. For example, a typical food chain in a grassland might be grass (producer) → mouse (primary consumer) → snake (secondary consumer) → hawk (tertiary consumer). Actually, in many cases the food chains of the ecosystem overlap and interconnect, forming what ecologists call a food web. The final link in all food chains is made up of decomposers, those heterotrophs that break down dead organisms and organic wastes. A food chain in which the primary consumer feeds on living plants is called a grazing pathway; that in which the primary consumer feeds on dead plant matter is known as a detritus pathway. Both pathways are important in accounting for the energy budget of the ecosystem.

Processes

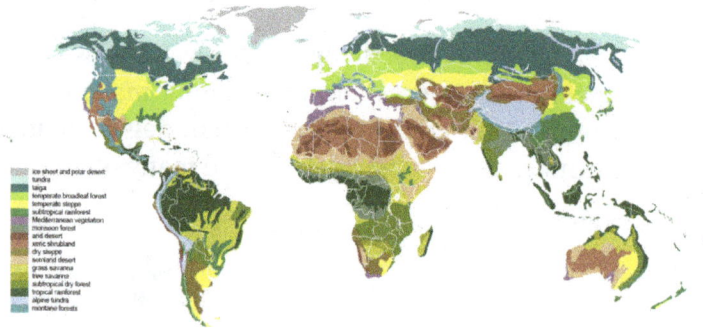

Biomes of the world

Flora of Baja California Desert, Cataviña region, Mexico

Rainforest ecosystems are rich in biodiversity. This is the Gambia River in Senegal's Niokolo-Koba National Park

External and Internal Factors

Ecosystems are controlled both by external and internal factors. External factors, also called state factors, control the overall structure of an ecosystem and the way things work within it, but are not themselves influenced by the ecosystem. The most important of these is climate. Climate determines the biome in which the ecosystem is embedded. Rainfall patterns and seasonal temperatures influence photosynthesis and thereby determine the amount of water and energy available to the ecosystem.

Parent material determines the nature of the soil in an ecosystem, and influences the supply of mineral nutrients. Topography also controls ecosystem processes by affecting things like microclimate, soil development and the movement of water through a system. For example, ecosystems can be quite different if situated in a small depression on the landscape, versus one present on an adjacent steep hillside.

Other external factors that play an important role in ecosystem functioning include time and potential biota. Similarly, the set of organisms that can potentially be present in an area can also significantly affect ecosystems. Ecosystems in similar environments that are located in different parts of the world can end up doing things very differently simply because they have different pools of species present. The introduction of non-native species can cause substantial shifts in ecosystem function.

Unlike external factors, internal factors in ecosystems not only control ecosystem processes but are also controlled by them. Consequently, they are often subject to feedback loops. While the resource inputs are generally controlled by external processes like climate and parent material, the availability of these resources within the ecosystem is controlled by internal factors like decomposition, root competition or shading. Other factors like disturbance, succession or the types of species present are also internal factors.

Primary Production

Primary production is the production of organic matter from inorganic carbon sources. This mainly occurs through photosynthesis. The energy incorporated through this process supports life on earth, while the carbon makes up much of the organic matter in living and dead biomass, soil carbon and fossil fuels. It also drives the carbon cycle, which influences global climate via the greenhouse effect.

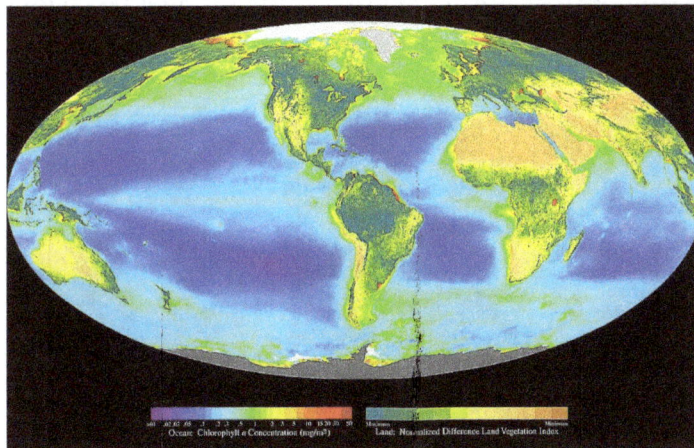

Global oceanic and terrestrial phototroph abundance. As an estimate of autotroph biomass, it is only a rough indicator of primary production potential and not an actual estimate of it

Through the process of photosynthesis, plants capture energy from light and use it to combine carbon dioxide and water to produce carbohydrates and oxygen. The photosynthesis carried out by all the plants in an ecosystem is called the gross primary production (GPP). About 48–60% of the GPP is consumed in plant respiration.

The remainder, that portion of GPP that is not used up by respiration, is known as the net primary production (NPP).

Energy Flow

Energy and carbon enter ecosystems through photosynthesis, are incorporated into living tissue, transferred to other organisms that feed on the living and dead plant matter, and eventually released through respiration.

The carbon and energy incorporated into plant tissues (net primary production) is either consumed by animals while the plant is alive, or it remains uneaten when the plant tissue dies and becomes detritus. In terrestrial ecosystems, roughly 90% of the net primary production ends up being broken down by decomposers. The remainder is either consumed by animals while still alive and enters the plant-based trophic system, or it is consumed after it has died, and enters the detritus-based trophic system.

In aquatic systems, the proportion of plant biomass that gets consumed by herbivores is much higher. In trophic systems photosynthetic organisms are the primary producers. The organisms that consume their tissues are called primary consumers or secondary producers—herbivores. Organisms which feed on microbes (bacteria and fungi) are termed microbivores. Animals that feed on primary consumers—carnivores—are secondary consumers. Each of these constitutes a trophic level.

The sequence of consumption—from plant to herbivore, to carnivore—forms a food chain. Real systems are much more complex than this—organisms will generally feed on more than one form of food, and may feed at more than one trophic level. Carnivores may capture some prey which are part of a plant-based trophic system and others that are part of a detritus-based trophic system (a bird that feeds both on herbivorous grasshoppers and earthworms, which consume detritus). Real systems, with all these complexities, form food webs rather than food chains.

Ecosystem Ecology

A hydrothermal vent is an ecosystem on the ocean floor. (The scale bar is 1 m.)

Ecosystem ecology studies "the flow of energy and materials through organisms and the physical environment". It seeks to understand the processes which govern the stocks of material and energy in ecosystems, and the flow of matter and energy through them. The study of ecosystems can cover 10 orders of magnitude, from the surface layers of rocks to the surface of the planet.

Decomposition

The carbon and nutrients in dead organic matter are broken down by a group of processes known as decomposition. This releases nutrients that can then be re-used for plant and microbial production and returns carbon dioxide to the atmosphere (or water) where it can be used for photosynthesis. In the absence of decomposition, the dead organic matter would accumulate in an ecosystem, and nutrients and atmospheric carbon dioxide would be depleted. Approximately 90% of terrestrial net primary production goes directly from plant to decomposer.

Decomposition processes can be separated into three categories—leaching, fragmentation and chemical alteration of dead material.

Leaching

As water moves through dead organic matter, it dissolves and carries with it the water-soluble components. These are then taken up by organisms in the soil, react with mineral soil, or are transported beyond the confines of the ecosystem (and are considered lost to it). Newly shed leaves and newly dead animals have high concentrations of water-soluble components and include sugars, amino acids and mineral nutrients. Leaching is more important in wet environments and much less important in dry ones.

Fragmentation

Fragmentation processes break organic material into smaller pieces, exposing new surfaces for colonization by microbes. Freshly shed leaf litter may be inaccessible due to an outer layer of cuticle or bark, and cell contents are protected by a cell wall. Newly dead animals may be covered by an exoskeleton. Fragmentation processes, which break through these protective layers, accelerate the rate of microbial decomposition. Animals fragment detritus as they hunt for food, as does passage through the gut. Freeze-thaw cycles and cycles of wetting and drying also fragment dead material.

Chemical Alteration

The chemical alteration of the dead organic matter is primarily achieved through bacterial and fungal action. Fungal hyphae produce enzymes which can break through the tough outer structures surrounding dead plant material. They also produce enzymes which break down lignin, which allows them access to both cell contents and to the nitrogen in the lignin. Fungi can transfer carbon and nitrogen through their hyphal networks and thus, unlike bacteria, are not dependent solely on locally available resources.

Decomposition Rates

Decomposition rates vary among ecosystems. The rate of decomposition is governed by three sets of factors—the physical environment (temperature, moisture, and soil properties), the quantity and quality of the dead material available to decomposers, and the nature of the microbial community itself. Temperature controls the rate of microbial respiration; the higher the temperature, the faster microbial decomposition occurs. It also affects soil moisture, which slows microbial growth and reduces leaching. Freeze-thaw cycles also affect decomposition—freezing temperatures kill soil microorganisms, which allows leaching to play a more important role in moving nutrients around. This can be especially important as the soil thaws in the spring, creating a pulse of nutrients which become available.

Decomposition rates are low under very wet or very dry conditions. Decomposition rates are highest in wet, moist conditions with adequate levels of oxygen. Wet soils tend to become deficient in oxygen (this is especially true in wetlands), which slows microbial growth. In dry soils, decomposition slows as well, but bacteria continue to grow (albeit at a slower rate) even after soils become too dry to support plant growth.

Nutrient Cycling

Ecosystems continually exchange energy and carbon with the wider environment. Mineral nutrients, on the other hand, are mostly cycled back and forth between plants, animals, microbes and the soil. Most nitrogen enters ecosystems through biological nitrogen fixation, is deposited through precipitation, dust, gases or is applied as fertilizer.

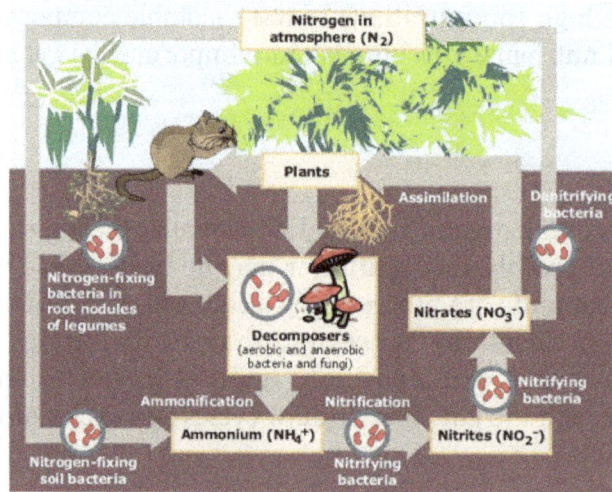

Biological nitrogen cycling

Nitrogen Cycle

Since most terrestrial ecosystems are nitrogen-limited, nitrogen cycling is an important control on ecosystem production.

Until modern times, nitrogen fixation was the major source of nitrogen for ecosystems. Nitrogen-fixing bacteria either live symbiotically with plants or live freely in the soil. The energetic cost is high for plants which support nitrogen-fixing symbionts—as much as 25% of gross primary production when measured in controlled conditions. Many members of the legume plant family support nitrogen-fixing symbionts. Some cyanobacteria are also capable of nitrogen fixation. These are phototrophs, which carry out photosynthesis. Like other nitrogen-fixing bacteria, they can either be free-living or have symbiotic relationships with plants. Other sources of nitrogen include acid deposition produced through the combustion of fossil fuels, ammonia gas which evaporates from agricultural fields which have had fertilizers applied to them, and dust. Anthropogenic nitrogen inputs account for about 80% of all nitrogen fluxes in ecosystems.

When plant tissues are shed or are eaten, the nitrogen in those tissues becomes available to animals and microbes. Microbial decomposition releases nitrogen compounds from dead organic matter in the soil, where plants, fungi, and bacteria compete for it. Some soil bacteria use organic nitrogen-containing compounds as a source of carbon, and release ammonium ions into the soil. This process is known as nitrogen mineralization. Others convert ammonium to nitrite and nitrate ions, a process known as nitrification. Nitric oxide and nitrous oxide are also produced during nitrification. Under nitrogen-rich and oxygen-poor conditions, nitrates and nitrites are converted to nitrogen gas, a process known as denitrification.

Other Nutrients

Other important nutrients include phosphorus, sulfur, calcium, potassium, magnesium and manganese. Phosphorus enters ecosystems through weathering. As ecosystems age this supply diminishes, making phosphorus-limitation more common in older landscapes (especially in the tropics). Calcium and sulfur are also produced by weathering, but acid deposition is an important source of sulfur in many ecosystems. Although magnesium and manganese are produced by weathering, exchanges between soil organic matter and living cells account for a significant portion of ecosystem fluxes. Potassium is primarily cycled between living cells and soil organic matter.

Function and Biodiversity

Loch Lomond in Scotland forms a relatively isolated ecosystem. The fish community of this lake has remained stable over a long period until a number of introductions restructured it's food web

Spiny forest at Ifaty, Madagascar, featuring various Adansonia (baobab) species, Alluaudia procera (Madagascar ocotillo) and other vegetation

Biodiversity plays an important role in ecosystem functioning. The reason for this is that ecosystem processes are driven by the number of species in an ecosystem, the exact nature of each individual species, and the relative abundance organisms within these species. Ecosystem processes are broad generalizations that actually take place through the actions of individual organisms. The nature of the organisms—the species, functional groups and trophic levels to which they belong—dictates the sorts of actions these individuals are capable of carrying out and the relative efficiency with which they do so.

Ecological theory suggests that in order to coexist, species must have some level of limiting similarity—they must be different from one another in some fundamental way, otherwise one species would competitively exclude the other. Despite this, the cumulative effect of additional species in an ecosystem is not linear—additional species may enhance nitrogen retention, for example, but beyond some level of species richness, additional species may have little additive effect.

The addition (or loss) of species which are ecologically similar to those already present in an eco-system tends to only have a small effect on ecosystem function. Ecologically distinct species, on the other hand, have a much larger effect. Similarly, dominant species have a large effect on ecosystem function, while rare species tend to have a small effect. Keystone species tend to have an effect on ecosystem function that is disproportionate to their abundance in an ecosystem. Similarly, an eco-system engineer is any organism that creates, significantly modifies, maintains or destroys a habitat.

Classification Methods

Classifying ecosystems into ecologically homogeneous units is an important step towards effective ecosystem management. There is no single, agreed-upon way to do this. A variety of systems exist, based on vegetation cover, remote sensing, and bioclimatic classification systems.

Ecological land classification is a cartographical delineation or regionalisation of distinct ecological areas, identified by their geology, topography, soils, vegetation, climate conditions, living species, habitats, water resources, and sometimes also anthropic factors.

Human Activities

Human activities are important in almost all ecosystems. Although humans exist and operate within ecosystems, their cumulative effects are large enough to influence external factors like climate.

Ecosystem Goods and Services

The High Peaks Wilderness Area in the 6,000,000-acre (2,400,000 ha)
Adirondack Park is an example of a diverse ecosystem

Ecosystems provide a variety of goods and services upon which people depend. Ecosystem goods include the "tangible, material products" of ecosystem processes such as food, construction material, medicinal plants. They also include less tangible items like tourism and recreation, and genes from wild plants and animals that can be used to improve domestic species.

Ecosystem services, on the other hand, are generally "improvements in the condition or location of things of value". These include things like the maintenance of hydrological cycles, cleaning air and water, the maintenance of oxygen in the atmosphere, crop pollination and even things like beauty, inspiration and opportunities for research. While ecosystem goods have traditionally been recognized as being the basis for things of economic value, ecosystem services tend to be taken for granted.

Ecosystem Management

When natural resource management is applied to whole ecosystems, rather than single species, it is termed ecosystem management. Although definitions of ecosystem management abound, there is a common set of principles which underlie these definitions. A fundamental principle is the long-term sustainability of the production of goods and services by the ecosystem; "intergenerational sustainability is a precondition for management, not an afterthought".

While ecosystem management can be used as part of a plan for wilderness conservation, it can also be used in intensively managed ecosystems.

Threats Caused by Humans

As human populations and per capita consumption grow, so do the resource demands imposed on ecosystems and the effects of the human ecological footprint. Natural resources are vulnerable and limited. The environmental impacts of anthropogenic actions are becoming more apparent. Problems for all ecosystems include: environmental pollution, climate change and biodiversity loss. For terrestrial ecosystems further threats include air pollution, soil degradation, and deforestation. For aquatic ecosystems threats include also unsustainable exploitation of marine resources (for example overfishing of certain species), marine pollution, microplastics pollution, water pollution, and building on coastal areas.

Society is increasingly becoming aware that ecosystem services are not only limited but also that they are threatened by human activities. The need to better consider long-term ecosystem health and its role in enabling human habitation and economic activity is urgent. To help inform decision-makers, many ecosystem services are being assigned economic values, often based on the cost of replacement with anthropogenic alternatives. The ongoing challenge of prescribing economic value to nature, for example through biodiversity banking, is prompting transdisciplinary shifts in how we recognize and manage the environment, social responsibility, business opportunities, and our future as a species.

Terrestrial Ecosystem

A terrestrial ecosystem is an ecosystem that exists on land, rather than on water. Such ecosystem is a community of organisms existing and living together on the land.

We can see this from the word terrestrial. 'Terrus' is Latin for land. Of course, water may be present in a terrestrial ecosystem. However, terrestrial ecosystems should primarily be situated on land.

Features of Terrestrial Ecosystem:

- Terrestrial ecosystems are ecosystems that exist on land.
- Etymologically, the word terrestrial comes from the word for land.

- Terrestrial ecosystems are distinct communities of organisms interacting and living together.

- There are many different types of terrestrial ecosystems.

- Terrestrial ecosystems can be distinguished from marine and fresh water ecosystems, which exist under water rather than on land.

Types of Terrestrial Ecosystem

There are many different types of terrestrial ecosystems in existence in the world. In order to understand terrestrial ecosystems better, we can divide them into several different categories, including:

Forest ecosystems: the animals, plants, insects and birds that live together in a forest. Forest ecosystems can take many forms, and they can exist on the mainland or near to the sea for example. Perhaps the most important forest ecosystems in the world are the rain forests, which account for a large part of our planet's biodiversity. Rain forests are usually humid and warm and the Amazon is a key example. Another forest ecosystem is the taiga, which is the name for the forest system that exists around the cold polar regions of our planet.

Desert ecosystems: the organisms that live in the world's deserts (be they sand deserts or ice deserts) are usually very hardy, as they have adapted to be able to live in very harsh conditions. A key example is the camel, which has a body that is able to store fluids so that the animal does not become dehydrated as it travels through the hot and sandy desert. Deserts do not have to be scorching hot (though this is very often our idea of a desert). Vast swathes of ice can also be referred to as deserts, as can the rocky regions of mountains or the cold, dark high pressure environment of the ocean floor. In fact, anywhere that is difficult to inhabit may be referred to as a desert ecosystem.

Grasslands: grassland or tundra is another great type of terrestrial ecosystem. Grassland can be home to migrating animals (such as buffalo) as well as to birds, predators, insects and humans. Grassland may be found all over the world and it may change vastly with the seasons or stay much the same throughout the year.

Six primary terrestrial ecosystems exist: tundra, taiga, temperate deciduous forest, tropical rain forest, grassland and desert.

• Taiga:

Taigas are cold-climate forests found in the northern latitudes. Taigas are the world's largest terrestrial ecosystem and account for about 29% of the Earth's forests. The largest taiga ecosystems are found in Canada and Russia. Taigas are known for their sub-arctic climate with extremely cold winters and mild summers. They primarily consist of coniferous trees, such as pines, although there are some other deciduous trees, such as spruce and elm, that have adapted to live in these areas that receive little direct sunlight for much of the year.Taigas are home to large herbivores, such as moose, elk, and bison, as well as omnivores, such as bears.

• Tundra:

The tundra ecosystems of the world are found primarily north of the Arctic Circle. They consist of short vegetation and essentially no trees. The soil is frozen and covered with permafrost for a large portion of the year. Caribou, polar bears, and musk ox are some of the notable species who call the tundra home.

• Temperate:

Temperate forests are the regions which have seasonal variation in climate i.e., the climate changes a lot from summer to winter. The annual rain fall is about 750- 2000 mm and soil is rich. Such types of forests are found in western and central Europe, Eastern Asia and eastern North America. These forests have deciduous trees (oaks, maples etc.) and coniferous trees (pines). These forests contain abundant micro-organ- isms, mammals (hares, deer, fares, coyotesetc). Birds (warblers, wood peckers, owls etc.) snakes, frogs, salamanders etc.

• Tropical Rain forests:

Tropical rain forests are special ecosystems which accommodate thousands of species of animals and plants. These are usually densely packed tall trees those form a ceiling from the sun above. The filing prevents the growth of smaller plants. The temperature remains almost same throughout the year. Such types of forests are found in Brazil of South America (Neotropic) and Central and West Africa. The area is always warm and muggy.

• Grasslands:

Grasslands are areas dominated by grasses. They occupy about 20% of the land on the earth surface. Grasslands occur in both in tropical and temperate regions where rainfall is not enough to support the growth of trees. Grasslands are found in areas having well defined hot and dry, warm and rainy seasons.

Place	Name of the Grassland
North America	Prairies
Eurasia (Europe and Asia)	Steppes
Africa	Savanna
South America	Pampas
India	Grassland, Savanna

• Deserts:

Desert are hot and low rain areas suffering from water shortage and high wind velocity. They show extremes of temperature. Globally deserts occupy about 1/7th of the earth's surface. Desert animals include shrew, fox, wood rats, rabbits, camels and goat are common mammals in desert Other prominent desert animals are, reptiles, and burrowing rodents insects. They adapt themselves to the dry weather conditions of desert area as stated in Adaptation topic.

Man Made Changes

All natural environments and ecosystems now have an unprecedented problem to deal with humanity. Humans have brought about profound changes in a few centuries which would otherwise be expected over thousands or millions of years. The full impact of these remains to be accurately estimated. Major human impacts on ecosystems include the following:

1. Habitat Destruction and Fragmentation

The most direct impact of humans on ecosystems is in their destruction or conversion. Clear-cutting (the cutting of all trees within a given forest area) will, obviously, destroy a forest ecosystem. Selective logging may also alter forest ecosystems in important ways. Fragmentation- the division of a once continuous ecosystem into a number of smaller patches- may disrupt ecological processes so that the remaining areas can no longer function as they once did.

2. Climate Change

It is now widely accepted that human activities are contributing to global warming, chiefly through the accumulation of "greenhouse" gases in the atmosphere. The impact of this is likely to increase in the future. As noted above, climate change is a natural feature of the Earth. Previously, however, its effects were mitigated as ecosystems could effectively "migrate" by moving latitude or altitude as the climate changed. Today, so much of the world's land surface has been appropriated by people that in many cases there is no such place for the remaining natural or semi-natural ecosystems to migrate to.

3. Pollution

Contamination of the natural environment through a range of pollutants- herbicides, pesticides, fertilizers, industrial effluents, and human waste products- is one of the most pernicious forms of impact on the natural environment. Pollutants are often invisible, and the effects of air pollution and water pollution may not be immediately obvious, although they can be devastating in the long run.

4. Introduced Species

Human beings have been responsible either deliberately or accidentally for altering the distribution of a vast range of animal and plant species. This includes not only domesticated animals and cultivated plants but pests such as rats, mice, and many insects and fungi. Species which become naturalized may have a devastating impact, through predation and competition, on natural ecosystems, particularly on islands where native species have evolved in isolation. For instance, foxes, rabbits, cane toads, feral cats, and even buffaloes and camels have wreaked havoc in many ecosystems in Australia. Plants such as the South American shrub Lantana have invaded natural forests in many tropical and subtropical islands, causing major changes to these ecosystems, while the African water hyacinth Eichhornia has similarly disrupted freshwater ecosystems in many of the warmer parts of the world.

5. Over-Harvest

Removal of excessive numbers of animals or plants from a system can cause major ecological changes. The most important example of this at present is the over-fishing of the world's oceans. Depletion of the great majority of accessible fish stocks is undoubtedly a cause of major change, although its long-term impact is difficult to assess.

Aquatic Ecosystem

Communities of plants and animals living in water are known as aquatic ecosystems. They are divided into two main groups. Freshwater ecosystems are found in water containing low concentrations of salts, from ponds to estuaries. Marine ecosystems are found in the saltwater of seas and oceans. Most of us are not far away from an aquatic ecosystem of some kind, whether it be in the ocean or a local pond.

The nature of an aquatic ecosystem is shaped, as on land, by the availability of food, oxygen, and the prevailing temperature. Added to this is salinity, which is the salt concentration of the water. Aquatic ecosystems in shallow waters, where there is plenty of sunlight, generally tend to be the most productive. Water pollution, generally coming from human activities, comprises the greatest pressure on aquatic ecosystems. For instance, fish can be killed by acid rain in lakes or lack of oxygen where excess nutrients have been dumped in an estuary.

Types

Marine

Marine ecosystems cover approximately 71% of the Earth's surface and contain approximately 97% of the planet's water. They generate 32% of the world's net primary production. They are distinguished from freshwater ecosystems by the presence of dissolved compounds, especially salts, in the water. Approximately 85% of the dissolved materials in seawater are sodium and chlorine. Seawater has an average salinity of 35 parts per thousand (ppt) of water. Actual salinity varies among different marine ecosystems.

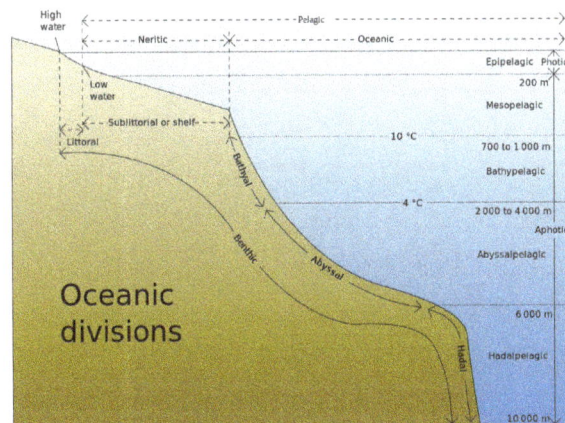

A classification of marine habitats.

Marine ecosystems can be divided into many zones depending upon water depth and shoreline features. The oceanic zone is the vast open part of the ocean where animals such as whales, sharks, and tuna live. The benthic zone consists of substrates below water where many invertebrates live. The intertidal zone is the area between high and low tides; in this figure it is termed the littoral zone. Other near-shore (neritic) zones can include estuaries, salt marshes, coral reefs, lagoons and mangrove swamps. In the deep water, hydrothermal vents may occur where chemosynthetic sulfur bacteria form the base of the food web.

Classes of organisms found in marine ecosystems include brown algae, dinoflagellates, corals, cephalopods, echinoderms, and sharks. Fishes caught in marine ecosystems are the biggest source of commercial foods obtained from wild populations.

Environmental problems concerning marine ecosystems include unsustainable exploitation of marine resources (for example overfishing of certain species), marine pollution, climate change, and building on coastal areas.

Freshwater

Freshwater ecosystems cover 0.78% of the Earth's surface and inhabit 0.009% of its total water. They generate nearly 3% of its net primary production. Freshwater ecosystems contain 41% of the world's known fish species.

Freshwater ecosystem.

There are three basic types of freshwater ecosystems:

- Lentic: slow moving water, including pools, ponds, and lakes.

- Lotic: faster moving water, for example streams and rivers.

- Wetlands: areas where the soil is saturated or inundated for at least part of the time.

Lentic

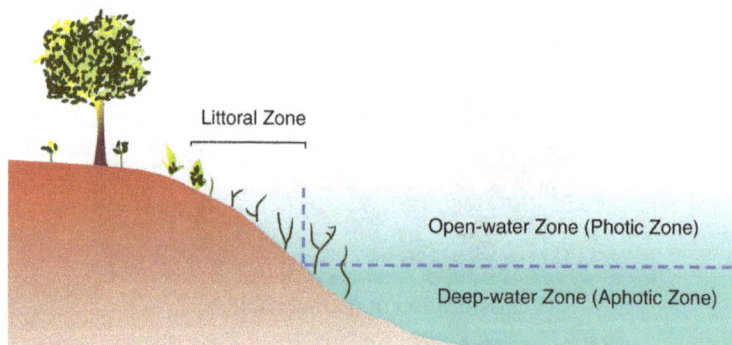

The three primary zones of a lake.

Lake ecosystems can be divided into zones. One common system divides lakes into three zones. The first, the littoral zone, is the shallow zone near the shore. This is where rooted wetland plants occur. The offshore is divided into two further zones, an open water zone and a deep water zone. In the open water zone (or photic zone) sunlight supports photosynthetic algae, and the species that feed upon them. In the deep water zone, sunlight is not available and the food web is based on detritus entering from the littoral and photic zones. Some systems use other names. The off shore areas may be called the pelagic zone, the photic zone may be called the limnetic zone and the aphotic zone may be called the profundal zone. Inland from the littoral zone one can also frequently identify a riparian zone which has plants still affected by the presence of the lake—this can include effects from windfalls, spring flooding, and winter ice damage. The production of the lake as a whole is the result of production from plants growing in the littoral zone, combined with production from plankton growing in the open water.

Wetlands can be part of the lentic system, as they form naturally along most lake shores, the width of the wetland and littoral zone being dependent upon the slope of the shoreline and the amount of natural change in water levels, within and among years. Often dead trees accumulate in this zone, either from windfalls on the shore or logs transported to the site during floods. This woody debris provides important habitat for fish and nesting birds, as well as protecting shorelines from erosion.

Two important subclasses of lakes are ponds, which typically are small lakes that intergrade with wetlands, and water reservoirs. Over long periods of time, lakes, or bays within them, may gradually become enriched by nutrients and slowly fill in with organic sediments, a process called succession. When humans use the watershed, the volumes of sediment entering the lake can accelerate this process. The addition of sediments and nutrients to a lake is known as eutrophication.

Ponds

Ponds are small bodies of freshwater with shallow and still water, marsh, and aquatic plants. They can be further divided into four zones: vegetation zone, open water, bottom mud and surface film. The size and depth of ponds often varies greatly with the time of year; many ponds are produced by spring flooding from rivers. Food webs are based both on free-floating algae and upon aquatic plants. There is usually a diverse array of aquatic life, with a few examples including algae, snails, fish, beetles, water bugs, frogs, turtles, otters and muskrats. Top predators may include large fish, herons, or alligators. Since fish are a major predator upon amphibian larvae, ponds that dry up each year, thereby killing resident fish, provide important refugia for amphibian breeding. Ponds that dry up completely each year are often known as vernal pools. Some ponds are produced by animal activity, including alligator holes and beaver ponds, and these add important diversity to landscapes.

Lotic

The major zones in river ecosystems are determined by the river bed's gradient or by the velocity of the current. Faster moving turbulent water typically contains greater concentrations of dissolved oxygen, which supports greater biodiversity than the slow moving water of pools. These distinctions form the basis for the division of rivers into upland and lowland rivers. The food base of streams within riparian forests is mostly derived from the trees, but wider streams and those that lack a canopy derive the majority of their food base from algae. Anadromous fish are also an important source of nutrients. Environmental threats to rivers include loss of water, dams, chemical pollution and introduced species. A dam produces negative effects that continue down the watershed. The most important negative effects are the reduction of spring flooding, which damages wetlands, and the retention of sediment, which leads to loss of deltaic wetlands.

Wetlands

Wetlands are dominated by vascular plants that have adapted to saturated soil. There are four main types of wetlands: swamp, marsh, fen and bog (both fens and bogs are types of mire).

Wetlands are the most productive natural ecosystems in the world because of the proximity of water and soil. Hence they support large numbers of plant and animal species. Due to their productivity, wetlands are often converted into dry land with dykes and drains and used for agricultural purposes. The construction of dykes, and dams, has negative consequences for individual wetlands and entire watersheds. Their closeness to lakes and rivers means that they are often developed for human settlement. Once settlements are constructed and protected by dykes, the settlements then become vulnerable to land subsidence and ever increasing risk of flooding. The Louisiana coast around New Orleans is a well-known example; the Danube Delta in Europe is another.

Functions

Aquatic ecosystems perform many important environmental functions. For example, they recycle nutrients, purify water, attenuate floods, recharge ground water and provide habitats for wildlife. Aquatic ecosystems are also used for human recreation, and are very important to the tourism industry, especially in coastal regions.

The health of an aquatic ecosystem is degraded when the ecosystem's ability to absorb a stress has been exceeded. A stress on an aquatic ecosystem can be a result of physical, chemical or biological alterations of the environment. Physical alterations include changes in water temperature, water flow and light availability. Chemical alterations include changes in the loading rates of biostimulatory nutrients, oxygen consuming materials, and toxins. Biological alterations include over-harvesting of commercial species and the introduction of exotic species. Human populations can impose excessive stresses on aquatic ecosystems. There are many examples of excessive stresses with negative consequences. Consider three. The environmental history of the Great Lakes of North America illustrates this problem, particularly how multiple stresses, such as water pollution, over-harvesting and invasive species can combine. The Norfolk Broadlands in England illustrate similar decline with pollution and invasive species. Lake Pontchartrain along the Gulf of Mexico illustrates the negative effects of different stresses including levee construction, logging of swamps, invasive species and salt water intrusion.

Abiotic Characteristics

An ecosystem is composed of biotic communities that are structured by biological interactions and abiotic environmental factors. Some of the important abiotic environmental factors of aquatic ecosystems include substrate type, water depth, nutrient levels, temperature, salinity, and flow. It is often difficult to determine the relative importance of these factors without rather large experiments. There may be complicated feedback loops. For example, sediment may determine the presence of aquatic plants, but aquatic plants may also trap sediment, and add to the sediment through peat.

The amount of dissolved oxygen in a water body is frequently the key substance in determining the extent and kinds of organic life in the water body. Fish need dissolved oxygen to survive, although their tolerance to low oxygen varies among species; in extreme cases of low oxygen some fish even resort to air gulping. Plants often have to produce aerenchyma, while the shape and size of leaves may also be altered. Conversely, oxygen is fatal to many kinds of anaerobic bacteria.

Nutrient levels are important in controlling the abundance of many species of algae. The relative abundance of nitrogen and phosphorus can in effect determine which species of algae come to dominate. Algae are a very important source of food for aquatic life, but at the same time, if they become over-abundant, they can cause declines in fish when they decay. Similar over-abundance of algae in coastal environments such as the Gulf of Mexico produces, upon decay, a hypoxic region of water known as a dead zone.

The salinity of the water body is also a determining factor in the kinds of species found in the water body. Organisms in marine ecosystems tolerate salinity, while many freshwater organisms are intolerant of salt. The degree of salinity in an estuary or delta is an important control upon the type of wetland (fresh, intermediate, or brackish), and the associated animal species. Dams built upstream may reduce spring flooding, and reduce sediment accretion, and may therefore lead to saltwater intrusion in coastal wetlands.

Freshwater used for irrigation purposes often absorbs levels of salt that are harmful to freshwater organisms.

Biotic Characteristics

The biotic characteristics are mainly determined by the organisms that occur. For example, wetland plants may produce dense canopies that cover large areas of sediment—or snails or geese may graze the vegetation leaving large mud flats. Aquatic environments have relatively low oxygen levels, forcing adaptation by the organisms found there. For example, many wetland plants must produce aerenchyma to carry oxygen to roots. Other biotic characteristics are more subtle and difficult to measure, such as the relative importance of competition, mutualism or predation. There are a growing number of cases where predation by coastal herbivores including snails, geese and mammals appears to be a dominant biotic factor.

Autotrophic Organisms

Autotrophic organisms are producers that generate organic compounds from inorganic material. Algae use solar energy to generate biomass from carbon dioxide and are possibly the most important autotrophic organisms in aquatic environments. The more shallow the water, the greater the biomass contribution from rooted and floating vascular plants. These two sources combine to produce the extraordinary production of estuaries and wetlands, as this autotrophic biomass is converted into fish, birds, amphibians and other aquatic species.

Chemosynthetic bacteria are found in benthic marine ecosystems. These organisms are able to feed on hydrogen sulfide in water that comes from volcanic vents. Great concentrations of animals that feed on these bacteria are found around volcanic vents. For example, there are giant tube worms (Riftia pachyptila) 1.5 m in length and clams (Calyptogena magnifica) 30 cm long.

Heterotrophic Organisms

Heterotrophic organisms consume autotrophic organisms and use the organic compounds in their bodies as energy sources and as raw materials to create their own biomass. Euryhaline organisms are salt tolerant and can survive in marine ecosystems, while stenohaline or salt intolerant species can only live in freshwater environments.

Marine Ecosystem

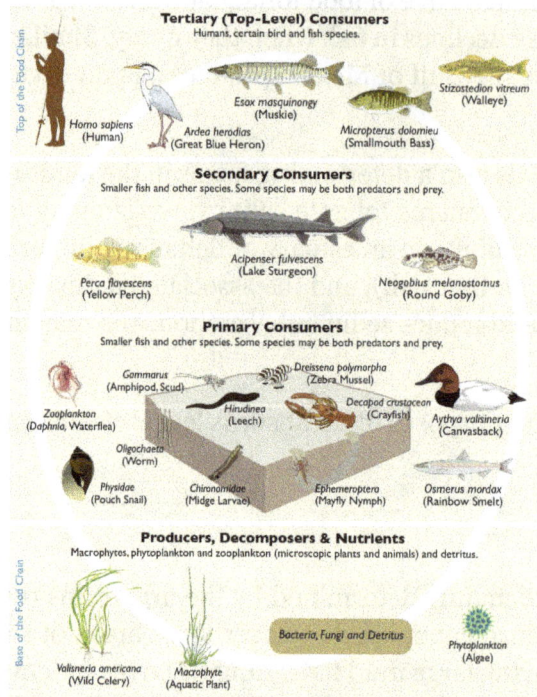

Marine ecosystem, complex of living organisms in the ocean environment

Marine waters cover two-thirds of the surface of the Earth. In some places the ocean is deeper than Mount Everest is high; for example, the Mariana Trench and the Tonga Trench in the western part of the Pacific Ocean reach depths in excess of 10,000 metres (32,800 feet). Within this ocean habitat live a wide variety of organisms that have evolved in response to various features of their environs.

Origins of Marine Life

The Earth formed approximately 4.5 billion years ago. As it cooled, water in the atmosphere condensed and the Earth was pummeled with torrential rains, which filled its great basins, forming seas. The primeval atmosphere and waters harboured the inorganic components hydrogen, methane, ammonia, and water. These substances are thought to have combined to form the first organic compounds when sparked by electrical discharges of lightning. Some of the earliest known organisms are cyanobacteria (formerly referred to as blue-green algae). Evidence of these early photosynthetic prokaryotes has been found in Australia in Precambrian marine sediments called stromatolites that are approximately 3 billion years old. Although the diversity of life-forms observed in modern oceans did not appear until much later, during the Precambrian (about 4.6 billion to 542 million years ago) many kinds of bacteria, algae, protozoa, and primitive metazoa evolved to exploit the early marine habitats of the world. During the Cambrian Period (about 542 million to 488 million years ago) a major radiation of life occurred in the oceans. Fossils of familiar organisms such as cnidaria (e.g., jellyfish), echinoderms (e.g., feather stars), precursors of the fishes (e.g., the protochordate Pikaia from the Burgess Shale of Canada), and other vertebrates are found in marine sediments of this age. The first fossil fishes are found in sediments from the Ordovician Period (about 488 million to 444 million years ago). Changes in the physical conditions of the ocean that are thought to have occurred in the Precambrian—an increase in the concentration of oxygen in seawater and a buildup of the ozone layer that reduced dangerous ultraviolet radiation—may have facilitated the increase and dispersal of living things.

The Marine Environment

Geography, Oceanography, and Topography

The shape of the oceans and seas of the world has changed significantly throughout the past 600 million years. According to the theory of plate tectonics, the crust of the Earth is made up of many dynamic plates. There are two types of plates—oceanic and continental—which float on the surface of the Earth's mantle, diverging, converging, or sliding against one another. When two plates diverge, magma from the mantle wells up and cools, forming new crust; when convergence occurs, one plate descends—i.e., is subducted—below the other and crust is resorbed into the mantle. Examples of both processes are observed in the marine environment. Oceanic crust is created along oceanic ridges or rift areas, which are vast undersea mountain ranges such as the Mid-Atlantic Ridge. Excess crust is reabsorbed along subduction zones, which usually are marked by deep-sea trenches such as the Kuril Trench off the coast of Japan.

The shape of the ocean also is altered as sea levels change. During ice ages a higher proportion of the waters of the Earth is bound in the polar ice caps, resulting in a relatively low sea level. When the polar ice caps melt during interglacial periods, the sea level rises. These changes in sea level cause great changes in the distribution of marine environments such as coral reefs. For example, during the last Pleistocene Ice Age the Great Barrier Reef did not exist as it does today; the continental shelf on which the reef now is found was above the high-tide mark.

Marine organisms are not distributed evenly throughout the oceans. Variations in characteristics of the marine environment create different habitats and influence what types of organisms will inhabit them. The availability of light, water depth, proximity to land, and topographic complexity all affect marine habitats.

The availability of light affects which organisms can inhabit a certain area of a marine ecosystem. The greater the depth of the water, the less light can penetrate until below a certain depth there is no light whatsoever. This area of inky darkness, which occupies the great bulk of the ocean, is called the aphotic zone. The illuminated region above it is called the photic zone, within which are distinguished the euphotic and disphotic zones. The euphotic zone is the layer closer to the surface that receives enough light for photosynthesis to occur. Beneath lies the disphotic zone, which is illuminated but so poorly that rates of respiration exceed those of photosynthesis. The actual depth of these zones depends on local conditions of cloud cover, water turbidity, and ocean surface. In general, the euphotic zone can extend to depths of 80 to 100 metres and the disphotic zone to depths of 80 to 700 metres. Marine organisms are particularly abundant in the photic zone, especially the euphotic portion; however, many organisms inhabit the aphotic zone and migrate vertically to the photic zone every night. Other organisms, such as the tripod fish and some species of sea cucumbers and brittle stars, remain in darkness all their lives.

Marine environments can be characterized broadly as a water, or pelagic, environment and a bottom, or benthic, environment. Within the pelagic environment the waters are divided into the neritic province, which includes the water above the continental shelf, and the oceanic province, which includes all the open waters beyond the continental shelf. The high nutrient levels of the neritic province—resulting from dissolved materials in riverine runoff—distinguish this province from the oceanic. The upper portion of both the neritic and oceanic waters—the epipelagic zone— is where photosynthesis occurs; it is roughly equivalent to the photic zone. Below this zone lie the mesopelagic, ranging between 200 and 1,000 metres, the bathypelagic, from 1,000 to 4,000 metres, and the abyssalpelagic, which encompasses the deepest parts of the oceans from 4,000 metres to the recesses of the deep-sea trenches.

The benthic environment also is divided into different zones. The supralittoral is above the high-tide mark and is usually not under water. The intertidal, or littoral, zone ranges from the high-tide mark (the maximum elevation of the tide) to the shallow, offshore waters. The sublittoral is the environment beyond the low-tide mark and is often used to refer to substrata of the continental shelf, which reaches depths of between 150 and 300 metres. Sediments of the continental shelf that influence marine organisms generally originate from the land, particularly in the form of riverine runoff, and include clay, silt, and sand. Beyond the continental shelf is the bathyal zone, which occurs at depths of 150 to 4,000 metres and includes the descending continental slope and rise. The abyssal zone (between 4,000 and 6,000 metres) represents a substantial portion of the oceans. The deepest region of the oceans (greater than 6,000 metres) is the hadal zone of the deep-sea trenches. Sediments of the deep sea primarily originate from a rain of dead marine organisms and their wastes.

Physical and Chemical Properties of Seawater

The physical and chemical properties of seawater vary according to latitude, depth, nearness to land, and input of fresh water. Approximately 3.5 percent of seawater is composed of dissolved compounds, while the other 96.5 percent is pure water. The chemical composition of seawater reflects such processes as erosion of rock and sediments, volcanic activity, gas exchange with the atmosphere, the metabolic and breakdown products of organisms, and rain. In addition to carbon, the nutrients essential for living organisms include nitrogen and phosphorus, which are minor constituents of seawater and thus are often limiting factors in organic cycles of the ocean. Concentrations of phosphorus and nitrogen are generally low in the photic zone because they are rapidly taken up by marine organisms. The highest concentrations of these nutrients generally are found below 500 metres, a result of the decay of organisms. Other important elements include silicon (used in the skeletons of radiolarians and diatoms) and calcium (essential in the skeletons of many organisms such as fish and corals).

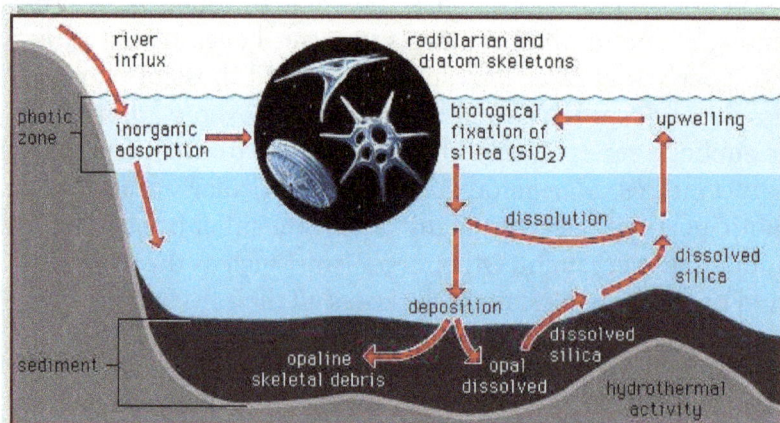

The chemical composition of the atmosphere also affects that of the ocean. For example, carbon dioxide is absorbed by the ocean and oxygen is released to the atmosphere through the activities of marine plants. The dumping of pollutants into the sea also can affect the chemical makeup of the ocean, contrary to earlier assumptions that, for example, toxins could be safely disposed of there.

The physical and chemical properties of seawater have a great effect on organisms, varying especially with the size of the creature. As an example, seawater is viscous to very small animals (less than 1 millimetre [0.039 inch] long) such as ciliates but not to large marine creatures such as tuna.

Marine organisms have evolved a wide variety of unique physiological and morphological features that allow them to live in the sea. Notothenid fishes in Antarctica are able to inhabit waters as cold as −2° C (28° F) because of proteins in their blood that act as antifreeze. Many organisms are able to achieve neutral buoyancy by secreting gas into internal chambers, as cephalopods do, or into swim bladders, as some fish do; other organisms use lipids, which are less dense than water, to achieve this effect. Some animals, especially those in the aphotic zone, generate light to attract prey. Animals in the disphotic zone such as hatchetfish produce light by means of organs called photophores to break up the silhouette of their bodies and avoid visual detection by predators. Many marine animals can detect vibrations or sound in the water over great distances by means of specialized organs. Certain fishes have lateral line systems, which they use to detect prey, and whales have a sound-producing organ called a melon with which they communicate. Tolerance to differences in salinity varies greatly: stenohaline organisms have a low tolerance to salinity changes, whereas euryhaline organisms, which are found in areas where river and sea meet (estuaries), are very tolerant of large changes in salinity. Euryhaline organisms are also very tolerant of changes in temperature. Animals that migrate between fresh water and salt water, such as salmon or eels, are capable of controlling their osmotic environment by active pumping or the retention of salts. Body architecture varies greatly in marine waters. The body shape of the cnidarian by-the-wind-sailor (Velella velella)—an animal that lives on the surface of the water (pleuston) and sails with the assistance of a modified flotation chamber—contrasts sharply with the sleek, elongated shape of the barracuda.

Ocean Currents

The movements of ocean waters are influenced by numerous factors, including the rotation of the Earth (which is responsible for the Coriolis effect), atmospheric circulation patterns that influence surface waters, and temperature and salinity gradients between the tropics and the polar regions (thermohaline circulation). The resultant patterns of circulation range from those that cover great areas, such as the North Subtropical Gyre, which follows a path thousands of kilometres long, to small-scale turbulences of less than one metre.

Marine organisms of all sizes are influenced by these patterns, which can determine the range of a species. For example, krill (Euphausia superba) are restricted to the Antarctic Circumpolar Current. Distribution patterns of both large and small pelagic organisms are affected as well. Mainstream currents such as the Gulf Stream and East Australian Current transport larvae great distances. As a result cold temperate coral reefs receive a tropical infusion when fish and invertebrate larvae from the tropics are relocated to high latitudes by these currents. The successful recruitment of eels to Europe depends on the strength of the Gulf Stream to transport them from spawning sites in the Caribbean. Areas where the ocean is affected by nearshore features, such as estuaries, or areas in which there is a vertical salinity gradient (halocline) often exhibit intense biological activity. In these environments, small organisms can become concentrated, providing a rich supply of food for other animals.

Marine Biota

Marine biota can be classified broadly into those organisms living in either the pelagic environment (plankton and nekton) or the benthic environment (benthos). Some organisms, however, are benthic in one stage of life and pelagic in another. Producers that synthesize organic molecules exist in both environments. Single-celled or multicelled plankton with photosynthetic pigments are the producers of the photic zone in the pelagic environment. Typical benthic producers are microalgae (e.g., diatoms), macroalgae (e.g., the kelp Macrocystis pyrifera), or sea grass (e.g., Zostera).

Plankton

Plankton are the numerous, primarily microscopic inhabitants of the pelagic environment. They are critical components of food chains in all marine environments because they provide nutrition for the nekton (e.g., crustaceans, fish, and squid) and benthos (e.g., sea squirts and sponges). They also exert a global effect on the biosphere because the balance of components of the Earth's atmosphere depends to a great extent on the photosynthetic activities of some plankton.

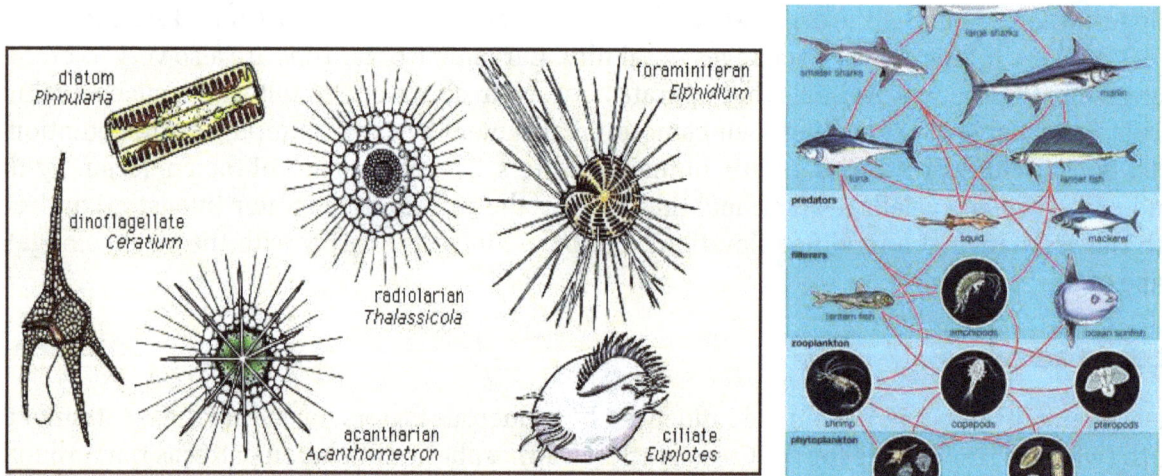

Figure: Representative plankton

The term plankton is derived from the Greek planktos, meaning wandering or drifting, an apt description of the way most plankton spend their existence, floating with the ocean's currents. Not all plankton, however, are unable to control their movements, and many forms depend on self-directed motions for their survival.

Plankton range in size from tiny microbes (1 micrometre [0.000039 inch] or less) to jellyfish whose gelatinous bell can reach up to 2 metres in width and whose tentacles can extend over 15 metres. However, most planktonic organisms, called plankters, are less than 1 millimetre (0.039 inch) long. These microbes thrive on nutrients in seawater and are often photosynthetic. The plankton include a wide variety of organisms such as algae, bacteria, protozoans, the larvae of some animals, and crustaceans. A large proportion of the plankton are protists—i.e., eukaryotic, predominantly single-celled organisms. Plankton can be broadly divided into phytoplankton, which are plants or plantlike protists; zooplankton, which are animals or animal-like protists; and microbes such as bacteria. Phytoplankton carry out photosynthesis and are the producers of the marine community; zooplankton are the heterotrophic consumers.

Diatoms and dinoflagellates (approximate range between 15 and 1,000 micrometres in length) are two highly diverse groups of photosynthetic protists that are important components of the plankton. Diatoms are the most abundant phytoplankton. While many dinoflagellates carry out photosynthesis, some also consume bacteria or algae. Other important groups of protists include flagellates, foraminiferans, radiolarians, acantharians, and ciliates. Many of these protists are important consumers and a food source for zooplankton.

Zooplankton, which are greater than 0.05 millimetre in size, are divided into two general categories: meroplankton, which spend only a part of their life cycle—usually the larval or juvenile stage—as plankton, and holoplankton, which exist as plankton all their lives. Many larval meroplankton in coastal, oceanic, and even freshwater environments (including sea urchins, intertidal snails, and crabs, lobsters, and fish) bear little or no resemblance to their adult forms. These larvae may exhibit features unique to the larval stage, such as the spectacular spiny armour on the larvae of certain crustaceans (e.g., Squilla), probably used to ward off predators.

Important holoplanktonic animals include such lobsterlike crustaceans as the copepods, cladocerans, and euphausids (krill), which are important components of the marine environment because they serve as food sources for fish and marine mammals. Gelatinous forms such as larvaceans, salps, and siphonophores graze on phytoplankton or other zooplankton. Some omnivorous zooplankton such as euphausids and some copepods consume both phytoplankton and zooplankton; their feeding behaviour changes according to the availability and type of prey. The grazing and predatory activity of some zooplankton can be so intense that measurable reductions in phytoplankton or zooplankton abundance (or biomass) occur. For example, when jellyfish occur in high concentration in enclosed seas, they may consume such large numbers of fish larvae as to greatly reduce fish populations.

The jellylike plankton are numerous and predatory. They secure their prey with stinging cells (nematocysts) or sticky cells (colloblasts of comb jellies). Large numbers of the Portuguese man-of-war (Physalia), with its conspicuous gas bladder, the by-the-wind-sailor (Velella velella), and the small blue disk-shaped Porpita porpita are propelled along the surface by the wind, and after strong onshore winds they may be found strewn on the beach. Beneath the surface, comb jellies often abound, as do siphonophores, salps, and scyphomedusae.

The pelagic environment was once thought to present few distinct habitats, in contrast to the array of niches within the benthic environment. Because of its apparent uniformity, the pelagic realm was understood to be distinguished simply by plankton of different sizes. Small-scale variations in the pelagic environment, however, have been discovered that affect biotic distributions. Living and dead matter form organic aggregates called marine snow to which members of the plankton community may adhere, producing patchiness in biotic distributions. Marine snow includes structures such as aggregates of cells and mucus as well as drifting macroalgae and other flotsam that range in size from 0.5 millimetre to 1 centimetre (although these aggregates can be as small as 0.05 millimetre and as large as 100 centimetres). Many types of microbes, phytoplankton, and zooplankton stick to marine snow, and some grazing copepods and predators will feed from the surface of these structures. Marine snow is extremely abundant at times, particularly after plankton blooms. Significant quantities of organic material from upper layers of the ocean may sink to the ocean floor as marine snow, providing an important source of food for bottom dwellers. Other structures that plankton respond to in the marine environment include aggregates of phytoplankton cells that form large rafts in tropical and temperate waters of the world (e.g., cells of Oscillatoria [Trichodesmium] erthraeus) and various types of seaweed (e.g., Sargassum, Phyllospora, Macrocystis) that detach from the seafloor and drift.

Nekton

Nekton are the active swimmers of the oceans and are often the best-known organisms of marine waters. Nekton are the top predators in most marine food chains. The distinction between nekton and plankton is not always sharp. As mentioned above, many large marine animals, such as marlin and tuna, spend the larval stage of their lives as plankton and their adult stage as large and active members of the nekton. Other organisms such as krill are referred to as both micronekton and macrozooplankton.

The vast majority of nekton are vertebrates (e.g., fishes, reptiles, and mammals), mollusks, and crustaceans. The most numerous group of nekton are the fishes, with approximately 16,000 species. Nekton are found at all depths and latitudes of marine waters. Whales, penguins, seals, and icefish abound in polar waters. Lantern fish (family Myctophidae) are common in the aphotic zone along with gulpers (Saccopharynx), whalefish (family Cetomimidae), seven-gilled sharks, and others. Nekton diversity is greatest in tropical waters, where in particular there are large numbers of fish species.

The largest animals on the Earth, the blue whales (Balaenoptera musculus), which grow to 25 to 30 metres long, are members of the nekton. These huge mammals and other baleen whales (order Mysticeti), which are distinguished by fine filtering plates in their mouths, feed on plankton and micronekton as do whale sharks (Rhinocodon typus), the largest fish in the world (usually 12 to 14 metres long, with some reaching 17 metres). The largest carnivores that consume large prey include the toothed whales (order Odontoceti—for example, the killer whales, Orcinus orca), great white sharks (Carcharodon carcharias), tiger sharks (Galeocerdo cuvier), black marlin (Makaira indica), bluefin tuna (Thunnus thynnus), and giant groupers (Epinephelus lanceolatus).

Nekton form the basis of important fisheries around the world. Vast schools of small anchovies, herring, and sardines generally account for one-quarter to one-third of the annual harvest from the ocean. Squid are also economically valuable nekton. Halibut, sole, and cod are demersal (i.e., bottom-dwelling) fish that are commercially important as food for humans. They are generally caught in continental shelf waters. Because pelagic nekton often abound in areas of upwelling where the waters are nutrient-rich, these regions also are major fishing areas.

Benthos

Organisms are abundant in surface sediments of the continental shelf and in deeper waters, with a great diversity found in or on sediments. In shallow waters, beds of seagrass provide a rich habitat for polychaete worms, crustaceans (e.g., amphipods), and fishes. On the surface of and within intertidal sediments most animal activities are influenced strongly by the state of the tide. On many sediments in the photic zone, however, the only photosynthetic organisms are microscopic benthic diatoms.

Benthic organisms can be classified according to size. The macrobenthos are those organisms larger than 1 millimetre. Those that eat organic material in sediments are called deposit feeders (e.g., holothurians, echinoids, gastropods), those that feed on the plankton above are the suspension feeders (e.g., bivalves, ophiuroids, crinoids), and those that consume other fauna in the benthic assemblage are predators (e.g., starfish, gastropods). Organisms between 0.1 and 1 millimetre constitute the meiobenthos. These larger microbes, which include foraminiferans, turbellarians, and polychaetes, frequently dominate benthic food chains, filling the roles of nutrient recycler, decomposer, primary producer, and predator. The microbenthos are those organisms smaller than 1 millimetre; they include diatoms, bacteria, and ciliates.

Organic matter is decomposed aerobically by bacteria near the surface of the sediment where oxygen is abundant. The consumption of oxygen at this level, however, deprives deeper layers of oxygen, and marine sediments below the surface layer are anaerobic. The thickness of the oxygenated layer varies according to grain size, which determines how permeable the sediment is to oxygen and the amount of organic matter it contains. As oxygen concentration diminishes, anaerobic processes come to dominate. The transition layer between oxygen-rich and oxygen-poor layers is called the redox discontinuity layer and appears as a gray layer above the black anaerobic layers. Organisms have evolved various ways of coping with the lack of oxygen. Some anaerobes release hydrogen sulfide, ammonia, and other toxic reduced ions through metabolic processes. The thiobiota, made up primarily of microorganisms, metabolize sulfur. Most organisms that live below the redox layer, however, have to create an aerobic environment for themselves. Burrowing animals generate a respiratory current along their burrow systems to oxygenate their dwelling places; the influx of oxygen must be constantly maintained because the surrounding anoxic layer quickly depletes the burrow of oxygen. Many bivalves (e.g., Mya arenaria) extend long siphons upward into oxygenated waters near the surface so that they can respire and feed while remaining sheltered from predation deep in the sediment. Many large mollusks use a muscular "foot" to dig with, and in some cases they use it to propel themselves away from predators such as starfish. The consequent "irrigation" of burrow systems can create oxygen and nutrient fluxes that stimulate the production of benthic producers (e.g., diatoms).

Not all benthic organisms live within the sediment; certain benthic assemblages live on a rocky substrate. Various phyla of algae—Rhodophyta (red), Chlorophyta (green), and Phaeophyta (brown)—are abundant and diverse in the photic zone on rocky substrata and are important producers. In intertidal regions algae are most abundant and largest near the low-tide mark. Ephemeral algae such as Ulva, Enteromorpha, and coralline algae cover a broad range of the intertidal. The mix of algae species found in any particular locale is dependent on latitude and also varies greatly according to wave exposure and the activity of grazers. For example, Ascophyllum spores cannot attach to rock in even a gentle ocean surge; as a result this plant is largely restricted to sheltered shores. The fastest-growing plant—adding as much as 1 metre per day to its length—is the giant kelp, Macrocystis pyrifera, which is found on subtidal rocky reefs. These plants, which may exceed 30 metres in length, characterize benthic habitats on many temperate reefs. Large laminarian and fucoid algae are also common on temperate rocky reefs, along with the encrusting (e.g., Lithothamnion) or short tufting forms (e.g., Pterocladia). Many algae on rocky reefs are harvested for food, fertilizer, and pharmaceuticals. Macroalgae are relatively rare on tropical reefs where corals abound, but Sargassum and a diverse assemblage of short filamentous and tufting algae are found, especially at the reef crest. Sessile and slow-moving invertebrates are common on reefs. In the intertidal and subtidal regions herbivorous gastropods and urchins abound and can have a great influence on the distribution of algae. Barnacles are common sessile animals in the intertidal. In the subtidal regions, sponges, ascidians, urchins, and anemones are particularly common where light levels drop and current speeds are high. Sessile assemblages of animals are often rich and diverse in caves and under boulders.

Reef-building coral polyps (Scleractinia) are organisms of the phylum Cnidaria that create a calcareous substrate upon which a diverse array of organisms live. Approximately 700 species of corals are found in the Pacific and Indian oceans and belong to genera such as Porites, Acropora, and Montipora. Some of the world's most complex ecosystems are found on coral reefs. Zooxanthellae are the photosynthetic, single-celled algae that live symbiotically within the tissue of corals and help to build the solid calcium carbonate matrix of the reef. Reef-building corals are found only in waters warmer than 18° C; warm temperatures are necessary, along with high light intensity, for the coral-algae complex to secrete calcium carbonate. Many tropical islands are composed entirely of hundreds of metres of coral built atop volcanic rock.

Links Between the Pelagic Environments and the Benthos

Considering the pelagic and benthic environments in isolation from each other should be done cautiously because the two are interlinked in many ways. For example, pelagic plankton are an important source of food for animals on soft or rocky bottoms. Suspension feeders such as anemones and barnacles filter living and dead particles from the surrounding water while detritus feeders graze on the accumulation of particulate material raining from the water column above. The molts of crustaceans, plankton feces, dead plankton, and marine snow all contribute to this rain of fallout from the pelagic environment to the ocean bottom. This fallout can be so intense in certain weather patterns—such as the El Niño condition—that benthic animals on soft bottoms are smothered and die. There also is variation in the rate of fallout of the plankton according to seasonal cycles of production. This variation can create seasonality in the abiotic zone where there is little or no variation in temperature or light. Plankton form marine sediments, and many types of fossilized protistan plankton, such as foraminiferans and coccoliths, are used to determine the age and origin of rocks.

Organisms of the Deep-sea Vents

Producers were discovered in the aphotic zone when exploration of the deep sea by submarine became common in the 1970s. Deep-sea hydrothermal vents now are known to be relatively common in areas of tectonic activity (e.g., spreading ridges). The vents are a nonphotosynthetic source of organic carbon available to organisms. A diversity of deep-sea organisms including mussels, large bivalve clams, and vestimentiferan worms are supported by bacteria that oxidize sulfur (sulfide) and derive chemical energy from the reaction. These organisms are referred to as chemoautotrophic, or chemosynthetic, as opposed to photosynthetic, organisms. Many of the species in the vent fauna have developed symbiotic relationships with chemoautotrophic bacteria, and as a consequence the megafauna are principally responsible for the primary production in the vent assemblage. The situation is analogous to that found on coral reefs where individual coral polyps have symbiotic relationships with zooxanthellae. In addition to symbiotic bacteria there is a rich assemblage of free-living bacteria around vents. For example, Beggiatoas-like bacteria often form conspicuous weblike mats on any hard surface; these mats have been shown to have chemoautotrophic metabolism. Large numbers of brachyuran (e.g., Bythograea) and galatheid crabs, large sea anemones (e.g., Actinostola callasi), copepods, other plankton, and some fish—especially the eelpout Thermarces cerberus—are found in association with vents.

Galatheid crabs and shrimp grazing on the bacterial filaments that grow on
the shells of the hydrothermal mussels covering the Northwest Eifuku volcano in the Mariana Arc region

Marine Ecosystem: Its Meaning and Types

An ecosystem can be defined as a community or system of organisms, all living together. This community is marked out as distinctive because it has certain defining characteristics.

But, what is a marine ecosystem? The word marine comes from the Latin word mare which means the sea. So, it follows from this that a marine ecosystem is an ecosystem based around the sea.

The sea is saline or salty. So, we can also add to this definition of a marine ecosystem the fact that this is a salt water ecosystem.

There are many seas and oceans all throughout the world. We might debate about whether they all form one giant ecosystem or whether they are all distinct ecosystems.

However, there are (to recap the above discussion) several things that we can say for definite about a marine ecosystem. That is, that it is:

- Based around the sea.

- A salt water ecosystem.

- Consisting of a community of organisms that live together.

- This community shares distinctive characteristics.

- Can vary depending on where the sea or ocean in question is located.

Types of Marine Ecosystem

There are various types of marine ecosystem, and understanding these different types can also help us to better understand the idea of a marine ecosystem as a whole. So, let us look at some of the key types.

1. Salt marshes: here, sea water saturates the land to create a unique environment where salt water creatures and plants can flourish.

2. Estuaries: where the river meets the sea, we have a fascinating tidal ecosystem which is home to both native and migrating birds and a wide variety of sea and river life.

3. The ocean floor: deep on the ocean floor there is little light but plenty of life. Crustaceans and other weird and wonderful deep sea creatures live here.

4. The broad ocean: right out to sea we can find whales, sharks, manatees and a wide variety of other wildlife.

5. The inter-tidal zone: right on the shore line, an ecosystem forms. Here, the types of creatures that we find in this ecosystem may well depend on whether it is high or low tide. This ecosystem may change vastly throughout the day as a result, but it is always a distinct and recognizable ecosystem.

6. Coral reefs: where living corals form a natural barrier, a new ecosystem can be located. Crustaceans, fish and other organisms base their lives around the coral. The Great Barrier Reef is a world famous example of this type of marine ecosystem.

7. Lagoons: where the sea spills in to a rocky valley, a static lagoon is created. In this calm world, many different types of birds, crustaceans, fish and other animals set up their homes, creating an interesting salt water ecosystem.

River Ecosystem

The ecology of the river refers to the relationships that living organisms have with each other and with their environment – the ecosystem. An ecosystem is the sum of interactions between plants, animals and microorganisms and between them and non-living physical and chemical components in a particular natural environment.

River ecosystems have:

- flowing water that is mostly unidirectional

- a state of continuous physical change

- many different (and changing) microhabitats

- variability in the flow rates of water

- plants and animals that have adapted to live within water flow conditions.

Kōtare or kingfisher

Kōtare or kingfishers that inhabit river areas eat small fish, insects and freshwater crayfish. If these river species decline, the kingfisher will move to another habitat.

Water Flow

Water flow is the main factor that makes river ecology different from other water ecosystems. This is known as a lotic (flowing water) system. The strength of water flow varies from torrential rapids to slow backwaters. The speed of water also varies and is subject to chaotic turbulence. Flow can be affected by sudden water input from snowmelt, rain and groundwater. Water flow can alter the shape of riverbeds through erosion and sedimentation, creating a variety of changing habitats.

The Waikato River is a lotic (flowing water) system

Substrate

The substrate is the surface on which the river organisms live. It may be inorganic, consisting of geological material from the catchment area such as boulders, pebbles, gravel, sand or silt, or it may be organic, including fine particles, leaves, wood, moss and plants. Substrate is generally not permanent and is subject to large changes during flooding events.

Light

Light provides energy for photosynthesis, which produces the primary food source for the river. It also provides refuges for prey species in the shadows it casts. The amount of light received in a flowing waterway is variable, for example, depending on whether it's a stream within a forest shaded by overhanging trees or a wide exposed river where the Sun has open access to its surface. Deep rivers tend to be more turbulent, and particles in the water increasingly weaken light penetration as depth increases.

Kaitiaki are working to restore and protect the health and wellbeing of the Waikato River

Temperature

Water temperature in rivers varies with the environment. Water can be heated or cooled through radiationat the surface and conduction to or from the air and surrounding substrate. Temperature differences can be significant between the surface and the bottom of deep, slow-moving rivers. Climate, shading and elevation all affect water temperature. Species living in these environments are called poikilotherms – their internal temperature varies to suit their environmental conditions.

Water Chemistry

The chemistry of the water varies from one river ecosystem to another. It is often determined by inputs from the surrounding environment or catchment area but can also be influenced by rain and the addition of pollution from human sources.

Oxygen is the most important chemical constituent of river systems – most organisms need it for survival. It enters the water mostly at the surface, but its solubility decreases as the water temperature increases. Fast, turbulent waters expose a wider water surface to the air and tend to have lower temperatures – achieving more oxygen input than slow backwaters. Oxygen is limited if water circulation is poor, animal activity is high or if there is a large amount of organic decay in the waterway.

Bacteria

Bacteria are present in large numbers in river waters. They play a significant role in energy recycling. Bacteria decompose organic material into inorganic compounds that can be used by plants and by other microbes.

Plants

Plants photosynthesise – converting light energy from the Sun into chemical energy that can be used to fuel organisms' activities.

Underwater plants: A variety of plants can be found growing within a river system.
Some plants are free-floating while others are rooted in areas of reduced current

Algae are the most significant source of primary food in most rivers or streams. Most float freely and are therefore unable to maintain large populations in fast-flowing water. They build up large numbers in slow-moving rivers or backwaters. Some algae species attach themselves to objects to avoid being washed away.

Plants are most successful in slower currents. Some plants such as mosses attach themselves to solid objects. Some plants are free-floating such as duckweed or water hyacinth. Others are rooted in areas of reduced current where sediment is found. Water currents provide oxygen and nutrients for plants. Plants protect animals from the current and predators and provide a food source.

Invertebrates

Invertebrates have no backbone or spinal column and include crayfish, snails, limpets, clams and mussels found in rivers. A large number of the invertebrates in river systems are insects. They can be found in almost every available habitat – on the water surface, on and under stones, in or below the substrate or adrift in the current. Some avoid high currents by living in the substrate area, while others have adapted by living on the sheltered downstream side of rocks. Invertebrates rely on the current to bring them food and oxygen. They are both consumers and prey in river systems.

Fish

Banded kōkopu are found in the Waikato River system. They are one of the species that make up whitebait

The ability of fish to live in a river system depends on their speed and duration of that speed – it takes enormous energy to swim against a current. This ability varies and is related to the area of habitat the fish may occupy in the river. Most fish tend to remain close to the bottom, the banks or behind obstacles, swimming in the current only to feed or change location. Some species never go into the current. Most river systems are typically connected to other lotic systems (springs, wetlands, waterways, streams, oceans), and many fish have life cycles that require stages in other systems. Eels, for example, move between freshwater and saltwater. Fish are important consumers and prey species.

Birds

A large number of birds also inhabit river ecosystems, but they are not tied to the water as fish are and spend some of their time in terrestrial habitats. Fish and water invertebrates are an important food source for water birds.

Ecosystem Management

Ecosystem management is an integrated approach to managing the health and diversity of natural systems to ensure the continuation of ecosystem goods and services for societal needs. Ecosystem goods and services are functions an ecosystem provides that are necessary for human growth and success. Some examples of these ecosystem goods and services include clean air, water and soil. Ecosystem management is a science-based approach that accounts for different ecosystem uses of air, land and water by humans while at the same time ensuring the health, resiliency and longevity of the ecosystem. Currently, water systems globally are stressed and continue to face increased pressure from development and human activity. To address this issue, ecosystem management creates a more holistic and integrated approach that takes into account elements needed to ensure healthy ecosystems and societies.

Ecosystem management helps to mitigate flooding in ways that are significantly different than engineered structures such as dams. Ecosystem management utilizes natural systems such a reforestation, designated riparian zones, and restoration of wetlands to mitigate against the damaging impacts of a flood. Natural solutions have the ability to slow down flood flows and retain flood waters in natural areas such as forests, wetlands, and floodplains. Furthermore, restorations of natural ecosystems have many positive feedback loops including climate stabilization, habitat restoration, and carbon and nitrogen storage.

Ecological Resilience

Ecological resilience, also called ecological robustness, the ability of an ecosystem to maintain its normal patterns of nutrient cycling and biomass production after being subjected to damage caused by an ecological disturbance. The term resilience is a term that is sometimes used interchangeably with robustness to describe the ability of a system to continue functioning amid and recover from a disturbance.

The resilience or robustness of ecological systems has been an important concept in ecology and natural history since the time of British naturalist Charles Darwin, who described the interdependencies between species as an "entangled bank" in his influential work On the Origin of Species (1859). Since then, the concept has come to hold special importance in the areas of environmental conservation and management. Its significance to the well-being of humans and human societies has also been recognized. The loss of an ecosystem's ability to recover from a disturbance—whether due to natural events such as hurricanes or volcanic eruptions or due to human influences such as overfishing and pollution—endangers the benefits (e.g., food, clean water, and aesthetics) that humans derive from that ecosystem.

However, resilience is not always a positive feature of a system. For example, an ecosystem may be locked in an undesirable state, such as in the case of a eutrophic lake, where an overabundance of nutrients results in hypoxia (depleted oxygen levels), which can lead to the demise of desirable fish species and the proliferation of undesirable pests.

Development of the Concept

In 1955 Canadian-born American ecologist Robert MacArthur proposed a measure of community stability that was related to the complexity of an ecosystem's food web. He stated that ecosystem stability increased as the number of interactions (complexity) between the different species within the ecosystem also increased. His collaborator, Australian theoretical physicist Robert May, later showed that communities of species that were more diverse and more complex were actually less able to maintain an exact stable numerical balance among species. This seemingly counterintuitive idea occurs because resilience or robustness at the level of the ecosystem is actually enhanced by a lack of rigidity at the level of its individual components (i.e., the populations or species within the ecosystem). This elasticity means that ecosystem properties, such as changes in nutrient flow or the number of species, are more resilient due to changes in species composition. For example, the disappearance of the American chestnut (Castanea dentata) in many forests in eastern North America due to chestnut blight has been largely compensated for by the expansion of oak (Quercus) and hickory (Carya) species, although there are certainly commercial consequences of this replacement.

In 1973 Canadian ecologist C.S. Holling wrote a paper that focused on the dichotomy between a type of resilience inherent in an engineered device (that is, the stability that comes from a machine designed to operate within a narrow range of expected circumstances) and the resilience that emphasizes an ecosystem's persistence as a particular ecosystem type (e.g., a forest as opposed to grassland), the latter being affected by substantially more factors than the former. Holling recognized the importance of the qualities that allowed a forest to persist as a functioning forest rather than its ability to harbour particular species at fixed levels or to maintain an arbitrary level of primary production. Holling's seminal paper brought heightened attention to the resilience of ecological systems and influenced other disciplines, such as economics and sociology. It has resonated in particular with the perspectives of individuals such as American biophysicist and geographer Jared Diamond, who is known for his examination of the conditions under which human societies developed, thrived, and collapsed.

Human Impacts

Resilience refers to ecosystem's stability and capability of tolerating disturbance and restoring itself. If the disturbance is of sufficient magnitude or duration, a threshold may be reached where the ecosystem undergoes a regime shift, possibly permanently. Sustainable use of environmental goods and services requires understanding and consideration of the resilience of the ecosystem and its limits. However, the elements which influence ecosystem resilience are complicated. For example, various elements such as the water cycle, fertility, biodiversity, plant diversity and climate, interact fiercely and affect different systems.

There are many areas where human activity impacts upon and is also dependent upon the resilience of terrestrial, aquatic and marine ecosystems. These include agriculture, deforestation, pollution, mining, recreation, overfishing, dumping of waste into the sea and climate change.

Agriculture

Agriculture can be seen as a significant example which the resilience of terrestrial ecosystems should be considered. The organic matter (elements carbon and nitrogen) in soil, which is supposed to be recharged by multiple plants, is the main source of nutrients for crop growth. At the same time, intensive agriculture practices in response to global food demand and shortages involves the removal of weeds and the application of fertilisers to increase food production. However, as a result of agricultural intensification and the application of herbicides to control weeds, fertilisers to accelerate and increase crop growth and pesticides to control insects, plant biodiversity is reduced as is the supply of organic matter to replenish soil nutrients and prevent run-off. This leads to a reduction in soil fertility and productivity. More sustainable agricultural practices would take into account and estimate the resilience of the land and monitor and balance the input and output of organic matter.

Deforestation

The term deforestation has a meaning that covers crossing the threshold of forest's resilience and losing its ability to return its originally stable state. To recover itself, a forest ecosystem needs suitable interactions among climate conditions and bio-actions, and enough area. In addition, generally, the resilience of a forest system allows recovery from a relatively small scale of damage (such as lightning or landslide) of up to 10 per cent of its area. The larger the scale of damage, the more difficult it is for the forest ecosystem to restore and maintain its balance.

Deforestation also decreases biodiversity of both plant and animal life and can lead to an alteration of the climatic conditions of an entire area. Deforestation can also lead to species extinction, which can have a domino effect particularly when keystone species are removed or when a significant number of species is removed and their ecological function is lost.

Climate Change

Climate resilience is generally defined as the capacity for a socio-ecological system to: (1) absorb stresses and maintain function in the face of external stresses imposed upon it by climate change and (2) adapt, reorganize, and evolve into more desirable configurations that improve the sustainability of the system, leaving it better prepared for future climate change impacts. Increasingly, climate change is threatening human communities around the world in a variety of ways such as rising sea levels, increasingly frequent large storms, tidal surges and flooding damage. One of the main results of climate change is rising sea water temperature which has a serious effect on coral reefs, through thermal-stress related coral bleaching. Between 1997-1998 the most significant worldwide coral bleaching event was recorded which corresponded with the El Nino Southern Oscillation, with significant damage to the coral reefs of the Western Indian Ocean.

Overfishing

It has been estimated by the United Nations Food and Agriculture Organisation that over 70% of the world's fish stocks are either fully exploited or depleted which means overfishing threatens marine ecosystem resilience and this is mostly by rapid growth of fishing technology. One of the negative effects on marine ecosystems is that over the last half-century the stocks of coastal fish have had a huge reduction as a result of overfishing for its economic benefits. Blue fin tuna is at particular risk of extinction. Depletion of fish stocks results in lowered biodiversity and consequently imbalance in the food chain, and increased vulnerability to disease.

In addition to overfishing, coastal communities are suffering the impacts of growing numbers of large commercial fishing vessels in causing reductions of small local fishing fleets. Many local lowland rivers which are sources of fresh water have become degraded because of the inflows of pollutants and sediments.

Dumping of Waste into the Sea

Dumping both depends upon ecosystem resilience whilst threatening it. Dumping of sewage and other contaminants into the ocean is often undertaken for the dispersive nature of the oceans and adaptive nature and ability for marine life to process the marine debris and contaminants. However, waste dumping threatens marine ecosystems by poisoning marine life and eutrophication.

Poisoning Marine Life

According to the International Maritime Organisation oil spills can have serious effects on marine life. The OILPOL Convention recognized that most oil pollution resulted from routine shipboard operations such as the cleaning of cargo tanks. In the 1950s, the normal practice was simply to wash the tanks out with water and then pump the resulting mixture of oil and water into the sea. OILPOL 54 prohibited the dumping of oily wastes within a certain distance from land and in 'special areas' where the danger to the environment was especially acute. In 1962 the limits were extended by means of an amendment adopted at a conference organized by IMO. Meanwhile, IMO in 1965 set up a Subcommittee on Oil Pollution, under the auspices of its Maritime Safety committee, to address oil pollution issues.

The threat of oil spills to marine life is recognised by those likely to be responsible for the pollution, such as the International Tanker Owners Pollution Federation.

The marine ecosystem is highly complex and natural fluctuations in species composition, abundance and distribution are a basic feature of its normal function. The extent of damage can therefore be difficult to detect against this background variability. Nevertheless, the key to understanding damage and its importance is whether spill effects result in a downturn in breeding success, productivity, diversity and the overall functioning of the system. Spills are not the only pressure on marine habitats; chronic urban and industrial contamination or the exploitation of the resources they provide are also serious threats.

Eutrophication and Algal Blooms

The Woods Hole Oceanographic Institution calls nutrient pollution the most widespread, chronic environmental problem in the coastal ocean. The discharges of nitrogen, phosphorus, and other nutrients come from agriculture, waste disposal, coastal development, and fossil fuel use. Once nutrient pollution reaches the coastal zone, it stimulates harmful overgrowths of algae, which can have direct toxic effects and ultimately result in low-oxygen conditions. Certain types of algae are toxic. Overgrowths of these algae result in harmful algal blooms, which are more colloquially referred to as "red tides" or "brown tides". Zooplankton eat the toxic algae and begin passing the toxins up the food chain, affecting edibles like clams, and ultimately working their way up to seabirds, marine mammals, and humans. The result can be illness and sometimes death.

Sustainable Development

There is increasing awareness that a greater understanding and emphasis of ecosystem resilience is required to reach the goal of sustainable development. A similar conclusion is drawn by Perman et al. who use resilience to describe one of 6 concepts of sustainability; "A sustainable state is one which satisfies minimum conditions for ecosystem resilience through time". Resilience science has been evolving over the past decade, expanding beyond ecology to reflect systems of thinking in fields such as economics and political science. And, as more and more people move into densely populated cities, using massive amounts of water, energy, and other resources, the need to combine these disciplines to consider the resilience of urban ecosystems and cities is of paramount importance.

Academic Perspectives

The interdependence of ecological and social systems has gained renewed recognition since the late 1990s by academics including Berkes and Folke and developed further in 2002 by Folke et al. As the concept of sustainable development has evolved beyond the 3 pillars of sustainable development to place greater political emphasis on economic development. This is a movement which causes wide concern in environmental and social forums and which Clive Hamilton describes as "the growth fetish".

The purpose of ecological resilience that is proposed is ultimately about averting our extinction as Walker cites Holling in his paper: "resilience is concerned with [measuring] the probabilities of extinction" . Becoming more apparent in academic writing is the significance of the environment and resilience in sustainable development. Folke et al state that the likelihood of sustaining development is raised by "Managing for resilience" whilst Perman et al. propose that safeguarding the environment to "deliver a set of services" should be a "necessary condition for an economy to be sustainable".

The Flaw of the Free Market

The challenge of applying the concept of ecological resilience to the context of sustainable development is that it sits at odds with conventional economic ideology and policy making. Resilience questions the free market model within which global markets operate. Inherent to the successful operation of a free market is specialisation which is required to achieve efficiency and increase productivity. This very act of specialisation weakens resilience by permitting systems to become accustomed to and dependent upon their prevailing conditions. In the event of unanticipated shocks; this dependency reduces the ability of the system to adapt to these changes. Correspondingly; Perman et al. note that; "Some economic activities appear to reduce resilience, so that the level of disturbance to which the ecosystem can be subjected to without parametric change taking place is reduced".

Moving beyond Sustainable Development

Berkes and Folke table a set of principles to assist with "building resilience and sustainability" which consolidate approaches of adaptive management, local knowledge-based management practices and conditions for institutional learning and self-organisation.

More recently, it has been suggested by Andrea Ross that the concept of sustainable development is no longer adequate in assisting policy development fit for today's global challenges and objectives. This is because the concept of sustainable development is "based on weak sustainability" which doesn't take account of the reality of "limits to earth's resilience". Ross draws on the impact of climate change on the global agenda as a fundamental factor in the "shift towards ecological sustainability" as an alternative approach to that of sustainable development.

In Environmental Policy

Scientific research associated with resilience is beginning to play a role in influencing policy-making and subsequent environmental decision making.

This occurs in a number of ways:

- Observed resilience within specific ecosystems drives management practice. When resilience is observed to be low, or impact seems to be reaching the threshold, management response can be to alter human behavior to result in less adverse impact to the ecosystem.

- Ecosystem resilience impacts upon the way that development is permitted/environmental decision making is undertaken, similar to the way that existing ecosystem health impacts upon what development is permitted. For instance, remnant vegetation in the states of Queensland and New South Wales are classified in terms of ecosystem health and abundance. Any impact that development has upon threatened ecosystems must consider the health and resilience of these ecosystems. This is governed by the Threatened Species Conservation Act 1995 in New South Wales and the Vegetation Management Act 1999 in Queensland.

- International level initiatives aim at improving socio-ecological resilience worldwide through the cooperation and contributions of scientific and other experts. An example of such an initiative is the Millennium Ecosystem Assessment whose objective is "to assess the consequences of ecosystem change for human well-being and the scientific basis for action needed to enhance the conservation and sustainable use of those systems and their contribution to human well-being". Similarly, the United Nations Environment Programme aim is "to provide leadership and encourage partnership in caring for the environment by inspiring, informing, and enabling nations and peoples to improve their quality of life without compromising that of future generations.

Environmental Management in Legislation

Ecological resilience and the thresholds by which resilience is defined are closely interrelated in the way that they influence environmental policy-making, legislation and subsequently environmental management. The ability of ecosystems to recover from certain levels of environmental impact is not explicitly noted in legislation, however, because of ecosystem resilience, some levels of environmental impact associated with development are made permissible by environmental policy-making and ensuing legislation.

Some examples of the consideration of ecosystem resilience within legislation include:

- Environmental Planning and Assessment Act 1979 (NSW) – A key goal of the Environmental Assessment procedure is to determine whether proposed development will have a significant impact upon ecosystems.

- Protection of the Environment (Operations) Act 1997 (NSW) – Pollution control is dependent upon keeping levels of pollutants emitted by industrial and other human activities below levels which would be harmful to the environment and its ecosystems. Environmental protection licenses are administered to maintain the environmental objectives of the POEO Act and breaches of license conditions can attract heavy penalties and in some cases criminal convictions.

- Threatened Species Conservation Act 1995 (NSW) – This Act seeks to protect threatened species while balancing it with development.

Resilience and the Development of Management Tools

Ecological resilience or robustness has also become central to conservation practices and ecosystem management, particularly as the latter has shifted its attention to the importance of ecosystem services. Such services include the provision of food, fuel, and natural products (e.g., substances for pharmaceutical development); the mediation of climate; the removal of toxic materials from environmental reservoirs; and the aesthetic enjoyment that humans derive from the natural world. Although many species retain importance within the framework of ecosystem services, much of the focus of conservation has moved from individual species to the maintenance of the ecosystem as a whole, especially its ability to retain its structure and rate of productivity.

Many lakes, for example, are managed to remain oligotrophic (relatively nutrient poor), with ample oxygen to support species such as lake trout, rather than managed to retain excess nutrients and algae. In addition, many terrestrial dryland ecosystems are managed to keep a richly vegetated area from undergoing desertification. Ecologists continue to look for ways to manage forests, such as those in Africa, to resist the transformation into a savanna through periods of extended drought or frequent wildfire episodes. Furthermore, in the ocean, where individual fish species have long been the subject of regulation, there is growing recognition of the need to expand efforts to manage large areas as integrated ecosystems.

Predicting the onset of disturbances such as eutrophication, desertification, and the collapse of fisheries has become an important component of ecosystem management. A greater emphasis on the identification of early-warning indicators, such as statistical fluctuations or correlations, has emerged. In particular, the ideas and techniques are being applied to medicine (such as in the onset of migraines or cardiac problems), research into climate change, and the operation of financial systems and markets. These indicators might serve as aids to management, much the same way that the detection of swarms of small earthquakes near a fault or an active volcano may portend the arrival of a larger seismic or eruptive event in the near future.

Equally important is the identification of the system's structural features that might impede the risk of systemic collapse or endow a system with the ability to recover from a disturbance. In ecological systems, ecologists might consider the diversity and heterogeneity among individual components (such as whole species, populations, or individual organisms) and landscape features within an ecosystem. Forest managers, for example, try to prevent the spread of wildfires throughout a forest by building firebreaks that follow changes in the landscape, such as those that separate one patch of trees from another. In addition, redundancy (niche overlap between species) and modularity (the interconnectedness of a system's components) are considered to be important factors that determine an ecosystem's resilience.

Ecosystem Services

Ecosystem services are the processes by which the environment produces resources utilised by humans such as clean air, water, food and materials. Ecosystem services can be defined in various ways.

Ecosystem services can be categorized in four main types:

- Provisioning services are the products obtained from ecosystems such as food, fresh water, wood, fiber, genetic resources and medicines.

- Regulating services are defined as the benefits obtained from the regulation of ecosystem processes such as climate regulation, natural hazard regulation, water purification and waste management, pollination or pest control.

- Habitat services highlight the importance of ecosystems to provide habitat for migratory species and to maintain the viability of gene-pools.

- Cultural services include non-material benefits that people obtain from ecosystems such as spiritual enrichment, intellectual development, recreation and aesthetic values.

Provisioning Services

Provisioning Services are ecosystem services that describe the material or energy outputs from ecosystems. They include food, water and other resources.

Food: Ecosystems provide the conditions for growing food. Food comes principally from managed agro-ecosystems but marine and freshwater systems or forests also provide food for human consumption. Wild foods from forests are often underestimated.

	Raw materials: Ecosystems provide a great diversity of materials for construction and fuel including wood, biofuels and plant oils that are directly derived from wild and cultivated plant species.
	Fresh water: Ecosystems play a vital role in the global hydrological cycle, as they regulate the flow and purification of water. Vegetation and forests influence the quantity of water available locally.
	Medicinal resources: Ecosystems and biodiversity provide many plants used as traditional medicines as well as providing the raw materials for the pharmaceutical industry. All ecosystems are a potential source of medicinal resources.

Regulating Services

Regulating Services are the services that ecosystems provide by acting as regulators eg. regulating the quality of air and soil or by providing flood and disease control.

	Local climate and air quality: Trees provide shade whilst forests influence rainfall and water availability both locally and regionally. Trees or other plants also play an important role in regulating air quality by removing pollutants from the atmosphere.
	Carbon sequestration and storage: Ecosystems regulate the global climate by storing and sequestering greenhouse gases. As trees and plants grow, they remove carbon dioxide from the atmosphere and effectively lock it away in their tissues. In this way forest ecosystems are carbon stores. Biodiversity also plays an important role by improving the capacity of ecosystems to adapt to the effects of climate change.
	Moderation of extreme events: Extreme weather events or natural hazards include floods, storms, tsunamis, avalanches and landslides. Ecosystems and living organisms create buffers against natural disasters, thereby preventing possible damage. For example, wetlands can soak up flood water whilst trees can stabilize slopes. Coral reefs and mangroves help protect coastlines from storm damage.
	Waste-water treatment: Ecosystems such as wetlands filter both human and animal waste and act as a natural buffer to the surrounding environment. Through the biological activity of microorganisms in the soil, most waste is broken down. Thereby pathogens (disease causing microbes) are eliminated, and the level of nutrients and pollution is reduced.
	Erosion prevention and maintenance of soil fertility: Soil erosion is a key factor in the process of land degradation and desertification. Vegetation cover provides a vital regulating service by preventing soil erosion. Soil fertility is essential for plant growth and agriculture and well functioning ecosystems supply the soil with nutrients required to support plant growth.
	Pollination: Insects and wind pollinate plants and trees which is essential for the development of fruits, vegetables and seeds. Animal pollination is an ecosystem service mainly provided by insects but also by some birds and bats. Some 87 out of the 115 leading global food crops depend upon animal pollination including important cash crops such as cocoa and coffee (Klein et al. 2007).

	Biological control: Ecosystems are important for regulating pests and vector borne diseases that attack plants, animals and people. Ecosystems regulate pests and diseases through the activities of predators and parasites. Birds, bats, flies, wasps, frogs and fungi all act as natural controls.

Habitat or Supporting Services

	Habitats for species: Habitats provide everything that an individual plant or animal needs to survive: food; water; and shelter. Each ecosystem provides different habitats that can be essential for a species' lifecycle. Migratory species including birds, fish, mammals and insects all depend upon different ecosystems during their movements.
	Maintenance of genetic diversity: Genetic diversity is the variety of genes between and within species populations. Genetic diversity distinguishes different breeds or races from each other thus providing the basis for locally well-adapted cultivars and a gene pool for further developing commercial crops and livestock. Some habitats have an exceptionally high number of species which makes them more genetically diverse than others and are known as 'biodiversity hotspots'.

Cultural Services

	Recreation and mental and physical health: Walking and playing sports in green space is not only a good form of physical exercise but also lets people relax. The role that green space plays in maintaining mental and physical health is increasingly being recognized, despite difficulties of measurement.
	Tourism: Ecosystems and biodiversity play an important role for many kinds of tourism which in turn provides considerable economic benefits and is a vital source of income for many countries. In 2008 global earnings from tourism summed up to US$ 944 billion. Cultural and eco-tourism can also educate people about the importance of biological diversity.
	Aesthetic appreciation and inspiration for culture, art and design: Language, knowledge and the natural environment have been intimately related throughout human history. Biodiversity, ecosystems and natural landscapes have been the source of inspiration for much of our art, culture and increasingly for science.
	Spiritual experience and sense of place: In many parts of the world natural features such as specific forests, caves or mountains are considered sacred or have a religious meaning. Nature is a common element of all major religions and traditional knowledge, and associated customs are important for creating a sense of belonging.

References

- Schowalter, Timothy Duane (2006). Insect ecology: an ecosystem approach (2(illustrated) ed.). Academic Press. p. 572. ISBN 978-0-12-088772-9. Retrieved 17 July 2010

- Lugo, A. E.; S.L. Brown; R. Dodson; T.S. Smith; H.H. Shugart (1999). "The Holdridge life zones of the conterminous United States in relation to ecosystem mapping" (PDF). Journal of Biogeography. 26 (5): 1025–1038. doi:10.1046/j.1365-2699.1999.00329.x

- Gullan, P.J.; Cranston, P.S. (2005). The insects: an outline of entomology (3 (illustrated, revised) ed.). Wiley-Blackwell. p. 505. ISBN 978-1-4051-1113-3. Retrieved 17 Jul 2010

- Kellogg, Charles (February 1933). "A Method for the Classification of Rural Lands for Assessment in Western North Dakota". The Journal of Land & Public Utility Economics. 9 (1): 12. JSTOR 3138756

- Speight, Martin R.; Hunter, Mark D.; Watt, Allan D. (1999). Ecology of insects: concepts and applications (4(Illustrated) ed.). Wiley-Blackwell. p. 350. ISBN 978-0-86542-745-7. Retrieved 2010-07-24

- Brown, Thomas C.; John C. Bergstrom; John B. Loomis (2007). "Defining, valuing and providing ecosystem goods and services" (PDF). Natural Resources Journal. 47 (2): 329–376. Archived from the original (PDF) on 2013-05-25

- HUFFAKER, CARL B. & GUTIERREZ, A. P. (1999). Ecological Entomology. 2nd Edition (illustrated). John Wiley and Sons. ISBN 0-471-24483-X, ISBN 978-0-471-24483-7.Limited preview on Google Books. Accessed on 09 Jan 2010

Chapter 2

Energy Flow in Ecosystems

In ecology, the flow of energy through a food chain is referred to as the energy flow. The food chain consists of the primary consumers or herbivores, carnivores or secondary consumers, tertiary consumers and decomposers. This chapter discusses in detail the process of nutrient cycle and energy flow in ecosystems. Some of the significant topics encompassed herein include food web, food chain, ecological pyramid, ecological efficiency, etc.

Ecosystem Ecology

Ecosystem ecology is the combined study of the physical and biological components of ecosystems. It focuses on how matter and energy flow through both organisms and the abiotic components of the environment.

Ecosystem ecology examines large-scale ecological issues, ones that often are framed in terms not of species but rather of measures such as biomass, energy flow, and nutrient cycling. Questions include how much carbon is absorbed from the atmosphere by terrestrial plants and marine phytoplankton during photosynthesis and how much of that is consumed by herbivores, the herbivores' predators, and so on up the food chain. Carbon is the basis of life, so these questions may be framed in terms of energy. How much food one has to eat each day, for instance, can be measured in terms of its dry weight or its calorie content. The same applies to measures of production for all the plants in an ecosystem or for different trophic levels of an ecosystem. A basic question in ecosystem ecology is how much production there is and what the factors are that affect it. Not surprisingly, warm, wet places such as rainforests produce more than extremely cold or dry places, but other factors are important. Nutrients are essential and may be in limited supply. The availability of phosphorus and nitrogen often determines productivity—it is the reason these substances are added to lawns and crops—and their availability is particularly important in aquatic systems. On the other hand, nutrients can represent too much of a good thing. Human activity has modified global ecosystems in ways that are increasing atmospheric carbon dioxide, a carbon source but also a greenhouse gas, and causing excessive runoff of fertilizers into rivers and then into the ocean, where it kills the species that live there.

Energy Flow

Living organisms can use energy in two forms radiant and fixed energy. Radiant energy is in the form of electromagnetic waves, such as light. Fixed energy is potential chemical energy bound in various organic substances which can be broken down in order to release their energy content.

Organisms that can fix radiant energy utilizing inorganic substances to produce organic molecules are called autotrophs. Organisms that cannot obtain energy from abiotic source but depend on energy-rich organic molecules synthesized by autotrophs are called heterotrophs. Those which obtain energy from living organisms are called consumers and those which obtain energy from dead organisms are called decomposers.

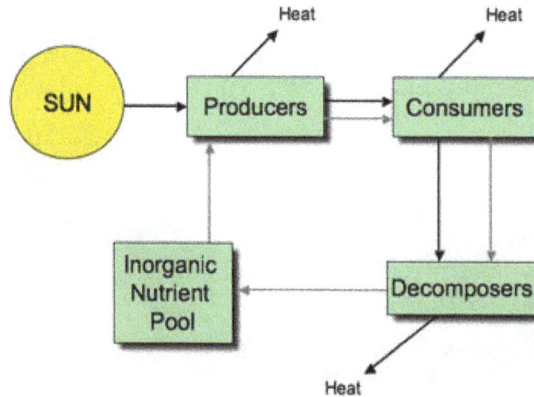

Flow of energy: Diffrent levels of ecosystem

When the light energy falls on the green surfaces of plants, a part of it is transformed into chemical energy which is stored in various organic products in the plants. When the herbivores consume plants as food and convert chemical energy accumulated in plant products into kinetic energy, degradation of energy will occur through its conversion into heat. When herbivores are consumed by carnivores of the first order (secondary consumers) further degradation will occur. Similarly, when primary carnivores are consumed by top carnivores, again energy will be degraded.

Trophic Level

The producers and consumers in ecosystem can be arranged into several feeding groups, each known as trophic level (feeding level). In any ecosystem, producers represent the first trophic level, herbivores present the second trophic level, primary carnivores represent the third trophic level and top carnivores represent the last level.

Food Chain

In the ecosystem, green plants alone are able to trap in solar energy and convert it into chemical energy. The chemical energy is locked up in the various organic compounds, such as carbohydrates, fats and proteins, present in the green plants. Since virtually all other living organisms depend upon green plants for their energy, the efficiency of plants in any given area in capturing solar energy sets the upper limit to long-term energy flow and biological activity in the community.

The food manufactured by the green plants is utilized by themselves and also by herbivores. Animals feed repeatedly. Herbivores fall prey to some carnivorous animals. In this way one form of life supports the other form. Thus, food from one trophic level reaches to the other trophic level and in this way a chain is established. This is known as food chain.

A food chain may be defined as the transfer of energy and nutrients through a succession of organisms through repeated process of eating and being eaten. In food chain initial link is a green plant or producer which produces chemical energy available to consumers. For example, marsh grass is consumed by grasshopper, the grasshopper is consumed by a bird and that bird is consumed by hawk.

Thus, a Food Chain is Formed which can be Written as Follows

Marsh grass → grasshopper → bird → hawk

Food chain in any ecosystem runs directly in which green plants are eaten by herbivores, herbivores are eaten by carnivores and carnivores are eaten by top carnivores. Man forms the terrestrial links of many food chains.

Food Chains are of Three Types

1. Grazing food chain

2. Parasitic food chain

3. Saprophytic or detritus food chain

1. Grazing Food Chain

The grazing food chain starts from green plants and from autotrophs it goes to herbivores (primary consumers) to primary carnivores (secondary consumers) and then to secondary carnivores (tertiary consumers) and so on. The gross production of a green plant in an ecosystem may meet three fates—it may be oxidized in respiration, it may be eaten by herbivorous animals and after the death and decay of producers it may be utilized by decomposers and converters and finally released into the environment. In herbivores the assimilated food can be stored as carbohydrates, proteins and fats, and transformed into much more complex organic molecules.

The energy for these transformations is supplied through respiration. As in autotrophs, the energy in herbivores also meets three routes respiration, decay of organic matter by microbes and consumption by the carnivores. Likewise, when the secondary carnivores or tertiary consumers eat primary carnivores, the total energy assimilated by primary carnivores or gross tertiary production follows the same course and its disposition into respiration, decay and further consumption by other carnivores is entirely similar to that of herbivores.

Thus, it is obvious that much of the energy flow in the grazing food chain can be described in terms of trophic levels as outlined below:

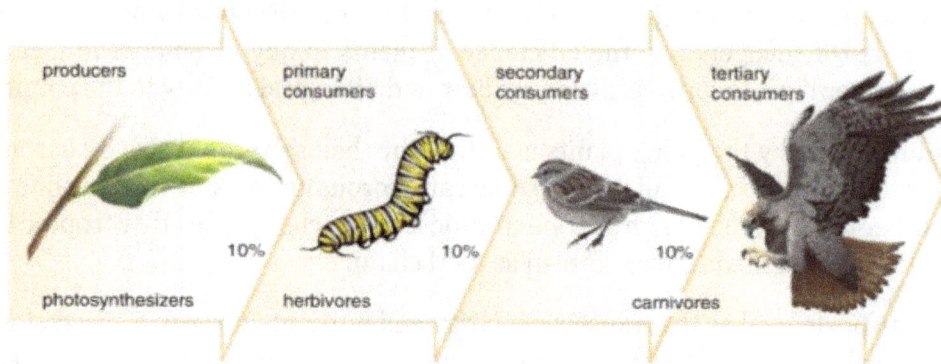

A schematic representation of grazing food chain showing input and losses of energy has been presented in figure

2. Parasitic Food Chain

It goes from large organisms to smaller ones without outright killing as in the case of predator.

3. Detritus Food Chain

The dead organic remains including metabolic wastes and exudates derived from grazing food chain are generally termed detritus. The energy contained in detritus is not lost in ecosystem as a whole, rather it serves as a source of energy for a group of organisms called detritivores that are separate from the grazing food chain. The food chain so formed is called detritus food chain.

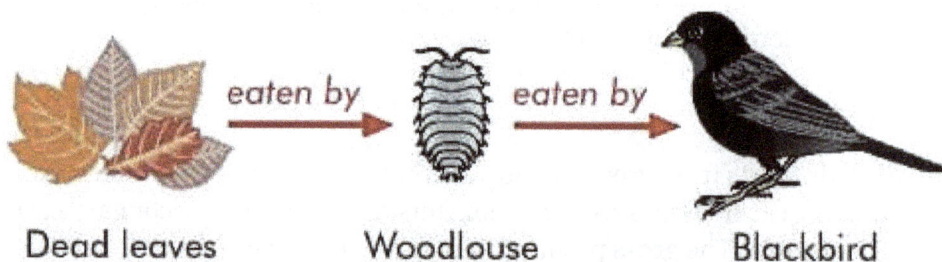

Dead leaves   eaten by   Woodlouse   eaten by   Blackbird

In some ecosystems more energy flows through the detritus food chain than through grazing food chain. In detritus food chain the energy flow remains as a continuous passage rather than as a stepwise flow between discrete entities. The organisms in the detritus food chain are many and include algae, fungi, bacteria, slime moulds, actinomycetes, protozoa, etc. Detritus organisms ingest pieces of partially decomposed organic matter, digest them partially and after extracting some of the chemical energy in the food to run their metabolism, excrete the remainder in the form of simpler organic molecules.

The waste from one organism can be immediately utilized by a second one which repeats the process. Gradually, the complex organic molecules present in the organic wastes or dead tissues are broken down to much simpler compounds, sometimes to carbon dioxide and water and all that are left are humus. In a normal environment the humus is quite stable and forms an essential part of the soil.

Food Web

Many food chains exist in an ecosystem, but as a matter of fact these food chains are not independent. In ecosystem, one organism does not depend wholly on another. The resources are shared specially at the beginning of the chain. The marsh plants are eaten by variety of insects, birds, mammals and fishes and some of the animals are eaten by several predators.

Similarly, in the food chain grass → mouse → snakes → owls, sometimes mice are not eaten by snakes but directly by owls. This type of interrelationship interlinks the individuals of the whole community. In this way, food chains become interlinked. A complex of interrelated food chains makes up a food web. Food web maintains the stability of the ecosystem. The greater the number of alternative pathways the more stable is the community of living things. Figure below illustrates a food web in ecosystem.

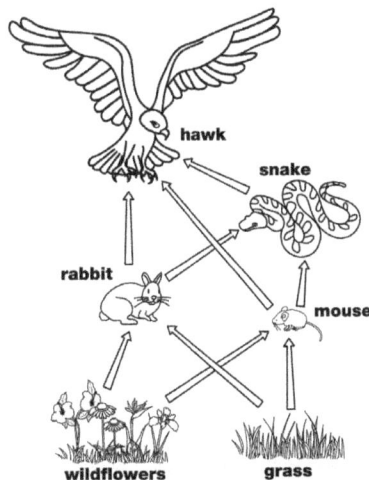

Figure: Food web in an ecosystem

Ecological Pyramid

The trophic structure of an ecosystem can be indicated by means of ecological pyramid. At each step in the food chain a considerable fraction of the potential energy is lost as heat. As a result, organisms in each trophic level pass on lesser energy to the next trophic level than they actually receive. This limits the number of steps in any food chain to 4 or 5. Longer the food chain the lesser energy is available for final members. Because of this tapering off of available energy in the food chain a pyramid is formed that is known as ecological pyramid. The higher the steps in the ecological pyramid the lower will be the number of individuals and the larger their size.

The idea of ecological pyramids was advanced by C.E. Elton (1927). There are different types of ecological pyramids. In each ecological pyramid, producer level forms the base and successive levels make up the apex. Three types of pyramidal relations may be found among the organisms at different levels in the ecosystem.

These are as Follows:

1.   Pyramid of numbers,

2.   Pyramid of biomass (biomass is the weight of living organisms), and

3.   Pyramid of energy.

1. Pyramid of Numbers

It depicts the numbers of individuals in producers and in different orders of consumers in an ecosystem. The base of pyramid is represented by producers which are the most abundant. In the successive levels of consumers, the number of organisms goes on decreasing rapidly until there are a few carnivores.

The pyramid of numbers of an ecosystem indicates that the producers are ingested in large numbers by smaller numbers of primary consumers. These primary consumers are eaten by relatively smaller number of secondary consumers and these secondary consumers, in turn, are consumed by only a few tertiary consumers.

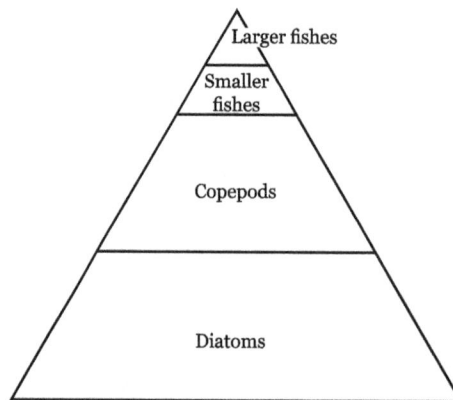

Figure: A Pyramid of numbers of a lake ecosystem

This type of pyramid is best presented by taking an example of Lake Ecosystem. In this type of pyramid the base trophic level is occupied by producer elements—algae, diatoms and other hydrophytes which are most abundant. At the second trophic level come the herbivores or zooplanktons which are lesser in number than producers.

The third trophic level is occupied by carnivores which are still smaller in number than the herbivores and the top is occupied by a few top carnivores. Thus, in the ecological pyramid of numbers there is a relative reduction in number of organisms and an increase in the size of body from base to apex of the pyramid. In parasitic food chain starting from tree, the pyramid of numbers will be inverted.

(A) Up-right Pyramids of numbers in a grassland and cultivated field
(B) Pyramids of numbers (inverted) of diseased tree (Parasitic ecosystem)

2. Pyramid of Biomass of Organisms

The living weights or biomass of the members of the food chain present at any one time form the pyramid of biomass of organisms. This indicates, by weight or other means of measuring materials, the total bulk of organisms or fixed energy present at one time. Pyramid of biomass indicates the decrease of biomass in each tropic level from base to apex, e.g., total biomass of producers is more than the total biomass of the herbivores.

Likewise, the total biomass of secondary consumers will be lesser than that of herbivores and so on. Since some energy and material are lost in each successive link, the total mass supported at each level is limited by the rate at which the energy is being stored below. This usually gives sloping pyramid for most of the communities in terrestrial and shallow water ecosystems. The pyramid of biomass in a pond ecosystem will be inverted as shown in figure below.

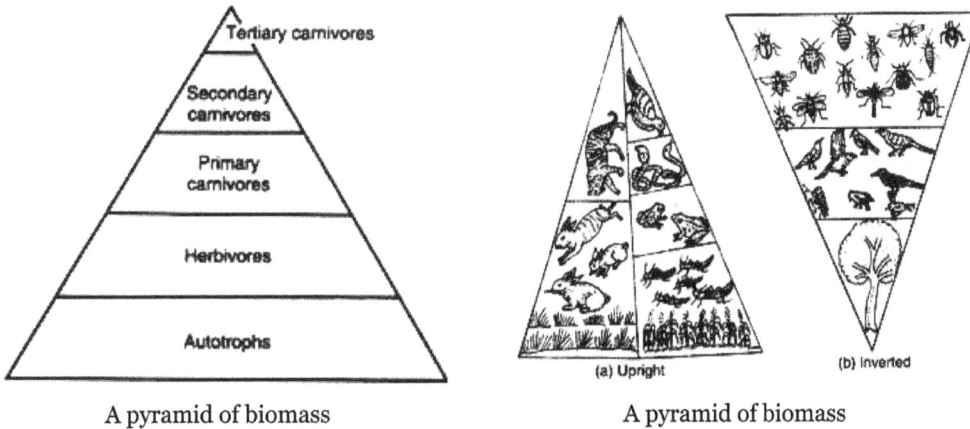

A pyramid of biomass   A pyramid of biomass

3. Pyramid of Energy

This depicts not only the amount of total energy utilized by the organisms at each trophic level of food chain but more important, the actual role of various organisms in transfer of energy. At the producer level the total energy will be much greater than the energy at the successive higher trophic level.

Some producer organisms may have small biomass but the total energy they assimilate and pass on to consumers may be greater than that of organisms with much larger biomass. Higher trophic levels are more efficient in energy utilization but much heat is lost in energy transfer. Energy loss by respiration also progressively increases from lower to higher trophic states.

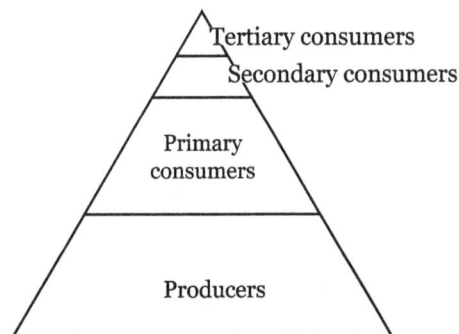

Figure: A pyramid of energy

In the energy flow process, two things become obvious. Firstly there is one way along which energy moves i.e. unidirectional flow of energy. Energy comes in the ecosystem from outside source i.e. sun. The energy captured by autotrophs does not go back to the sun, the energy that passes from autotrophs to herbivores does not revert back and as it moves progressively through the various trophic levels, it is no longer available to the previous levels.

Thus due to unidirectional flow of energy, the system would collapse if the supply from primary source, the sun is cut off. Secondly, there occurs a progressive decrease in energy level at each trophic level which is accounted largely by the energy dissipated as heat in metabolic activities.

Productivity

The relationship between the amount of energy accumulated and the amount of energy utilized within one trophic level of food chain has an important bearing on how much energy from one trophic level passes on to the next trophic level in the food chain. The ratio of output of energy to input of energy is referred to as ecological efficiency.

Different kinds of Efficiencies can be Measured by the Following Parameters

(i) Ingestion which indicates the quantity of food or energy taken by trophic level. This is also called exploitation efficiency.

(ii) Assimilation indicates the amount of food absorbed and fixed into energy rich organic substances which are stored or combined with other molecules to build complex molecules such as proteins, fats etc.

(iii) Respiration which indicates the energy lost in metabolism.

Primary Productivity

The fraction of fixed energy a trophic level passes on to the next trophic level is called production. Green plants fix solar energy and accumulate it in organic forms as chemical energy. Since it is the first and basic form of energy storage, the rate at which the energy accumulates in the green plants or producers is known as primary productivity.

Primary productivity is the rate at which energy is bound or organic material is created by photosynthesis per unit area of earth's surface per unit time. It is most often expressed as energy in calories / cm^2 / yr or dry organic matter in g / m^2 / yr (g/m^2 x 8.92 = lb / acre). The amount of organic matter present at a given time per unit area is called standing crop or biomass and as such productivity, which is a rate, is quite different from biomass or standing crop.

The standing crop is usually expressed as dry weight in g/m^2 or kg/m^2 or t/ha (metric tons) or 10^6g/hectare. Primary productivity is the result of photosynthesis by green plants including algae of different colours. Bacterial photosynthesis or chemosynthesis, although of small significance may also contribute to primary productivity. The total solar energy trapped in the food material by photosynthesis is referred to as gross primary productivity (G.P.P.).

A good fraction of gross primary production is utilized in respiration of green plants. The amount of energy bound in organic matter per unit area and time that is left after respiration in plants is net primary production (N.P.P.) or plant growth. Only the net primary productivity is available for harvest by man and other animals. Net productivity of energy = gross productivity—energy lost in respiration.

Secondary Productivity

The rates at which the heterotrophic organisms resynthesize the energy-yielding substances is termed as secondary productivity. Secondary productivities are the productivities of animals and saprobes in communities. The amount of energy stored in the tissues of consumers or heterotrophs is termed as net secondary production and the total plant material ingested by herbivores is grass secondary production. Total plant material ingested by herbivores minus the materials lost as faeces is equal to Ingested Secondary Production.

Environmental Factors Affecting the Production Processes in an Ecosystem are as Follows

1. Solar radiation and Temperature.

2. Moisture. Leaf water potential, soil moisture and precipitation fluctuation and transpiration.

3. Mineral nutrition. Uptake of minerals from the soil, rhizosphere effects, fire effects, salinity, heavy metals, nitrogen metabolism.

4. Biotic activities. Grazing, above ground herbivores, below ground herbivores, predators and parasites, diseases of primary producers.

5. Impact of human population. Pollutions of different sorts, ionizing radiations like atomic explosions, etc.

There are three Fundamental Concepts of Productivity

1. Standing crop

2. Materials removed

3. Production rate.

1. Standing crop

It is the abundance of the organisms existing in the area at any one time. It may be expressed in terms of number of individuals, as biomass of organisms, as energy content or in some other suitable terms. Measurement of standing crop reveals the concentration of individuals in the various populations of ecosystem.

2. The Materials Removed

The second concept of productivity is the materials removed from the area per unit time. It includes the yield to man, organisms removed from the ecosystem by migration, and the material withdrawn as organic deposit.

3. The Production Rate

The third concept of productivity is the production rate. It is the rate at which the growth processes are going forward within the area. The amount of material formed by each link in the food chain per unit of time per unit area or volume is the production rate.

All the three major groups of organisms—producers, consumers and reducers are the functional kingdoms of natural communities. The three represent major directions of evolution and are characterised by different modes of nutrition. Plants feed primarily by photosynthesis, animals feed primarily by ingesting food that is digested and absorbed in the alimentary canal and the saprobes feed by absorption and have need for an extensive surface of absorption. The principal kinds of organisms among saprobes are the unicellular bacteria, yeasts, chytrids or lower fungi and higher fungi with mycelial bodies.

In terrestrial communities as much as 90% of net primary production remains un-harvested and are utilized as dead tissue by saprobes and soil animals. The saprobes have a larger and more essential role than animals in degrading dead organic matter to inorganic forms and in such ecosystems, secondary production by reducers (decomposers) should exceed that by consumers, though the former is even more difficult to measure than the latter.

Biomass of decomposers with their microscopic cells and filaments embedded in food sources is also difficult to measure and that is small in relation to their productivity and significance for the ecosystem. Small masses of reducers degrade and transform larger masses of organic matter to inorganic remnants. In so doing decomposers disperse back to the environment the energy of photosynthesis accumulated in the organic compounds that are decomposed.

Thus they have a major role in the energy flow of ecosystems. A community or ecosystem, like an organism, is an open energy system. The continuous intake of energy in photosynthesis replaces the energy dissipated to environment by respiration and biological activity and the system does not run-down through the loss of free energy to maximum entropy.

If the amount of energy entrapped is greater than the energy dissipated, the pool of biologically useful energy of organic bonds increases. This results in increase of community biomass and consequently the community grows; such is the case in succession. If energy intake is lesser than energy dissipation, the community biomass will decrease and it must, in some sense, retrogress. If energy intake and loss are in balance, the pool of organic energy is in steady state; such is the case in climax communities.

Three Aspects of this Steady State may be Recognized

(i) The steady state of population of climax communities in which equal birth and death rates in population keep the number of individuals relatively constant,

(ii) The steady state of energy flow,

(iii)The steady state of the matter of community, where addition of material by photosynthesis and organic synthesis is balanced by loss of material through respiration and decomposition.

Methods of Measuring Primary Production

There are several parameters for measuring primary production and the methods of measuring primary production are based on those parameters.

The Methods are Discussed here as Under

1. Harvest Method

It involves removal of vegetation periodically and weighing the material. For measuring above ground production, the above ground plant parts are clipped at ground level, dried to constant weight at 80°C and weighed. The dry weight in g/m^2 /year gives the ground production. Below ground production is estimated by using frequent core sampling technique of Dahlman and Kucera (1965). It is expressed in terms of weight in gm per unit area per year. In terms of energy one gm dry weight of plant material contains 4 to 5 kcal.

The Limitations of Harvest Method are as Follows

(i) The amount of plant material consumed by herbivores and the food oxidized during respiration process of the plants is not accounted.

(ii) Root biomass is neglected.

(iii) Photosynthetic trans located to underground parts of plants are not known.

In spite of these limitations the method is used all over for measuring net assimilation rate (NAR) and relative growth rate (RGR).

2. Carbon Dioxide Assimilation Method

Utilization of CO_2 in photosynthesis or its liberation during respiration is measured by infrared gas analysis or by passing the gas through Baryta water $Ba(OH)_2$ and titrating the same. The CO_2 removed from incoming gas chamber is taken to be synthesized into organic matter by the green plants. Performing the experiment in light and dark chambers the net and gross production can be measured.

In the lighted chamber photosynthesis and respiration take place simultaneously and the CO_2 coming out from the chamber is the unused gas of the atmosphere plus gas from the respiration of plant parts. In the dark chamber all CO_2 is due to respiration.

Net production = Gross production—Respiration

3. Oxygen Production Method

In the aquatic vegetation CO_2 gas analysis method is not used but oxygen evolution method is generally used. The light and dark bottle technique is employed for measuring primary production of aquatic plant. In this method two bottles, one transparent and the other opaque are filled with water at a given depth of lake, closed, maintained at that depth for some time and then brought to laboratory for determination of oxygen content in the water. The decrease of oxygen in dark bottle is due to respiratory activity while increase of O_2 in light bottle is due to photosynthesis. The total increase of O_2 in light bottle plus the amount of O_2 decreased in dark bottle express gross productivity (O_2 value multiplied by 0.375 gives an equivalent of carbon assimilation). Recently, oxygen electrodes have been used for estimating oxygen content in water.

4. Chlorophyll Method

Gesner (1949) pointed out that the amount of chlorophyll/m² is almost limited to a narrow range of 0.1 to 3.0 gm regardless of the age of individuals or the species present therein. There is direct correlation between the amount of chlorophyll and dry matter production in different types of communities with varying light conditions.

The relation of total amount of chlorophyll to the photosynthetic rate is referred to as assimilation ratio or rate of production/gm chlorophyll. Total chlorophyll per unit area is greater in land plants as compared to that in aquatic plants. In marine ecosystem the rate of carbon assimilation is 3.7 g/ hr/g of chlorophyll. The relationship between area based chlorophyll and dry matter production in terrestrial ecosystems has been worked out by Japanese ecologists Argua and Monsi (1963).

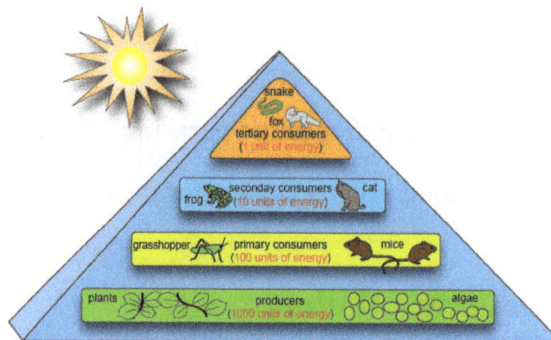

Ecosystem Engineer

Organisms impact the environment in which they live in in several ways. Some organisms are destructive to the ecosystem while others contribute positively to the development of the ecosystem. Organisms that create, modify, destroy or maintain a habitat in which they live or frequent are known as ecosystem engineers. These organisms can have a great impact on the species-richness and heterogeneity of the landscape of an area. Ecosystem Engineers maintain the health and stability of the environment they live in. However, since all organisms contribute to the modification of their environment in one way or the other, ecosystem engineer is only used to describe keystone species which plays a critical role in maintaining the ecological community and affects other organisms in the ecosystem.

Types of Ecosystem Engineers

Ecosystem engineers are divided into two broad categories: allogenic and autogenic engineers. Allogenic engineers modify the environment by mechanically transforming material, both living and non-living, from one form to another or various forms. Beavers, which are the original model for ecosystem engineers, alter their ecosystem extensively through the process of clear-cutting and damming. The addition of the dam changes the distribution and the abundance of the organisms in the area. By creating a shelter from leaves, a caterpillar also creates a shelter for another organism which may interact with it simultaneously or as a result of building the shelter. Birds that create holes in trees and wood to nest in also create homes for other organisms once they are through with them. Autogenic engineers, on the other hand, modify the environment on which they are in by modifying themselves. Trees are the best examples of autogenic engineers since as they grow, their trunks, leaves, and branches are used by other organisms as habitats including birds, insects, snakes, and other organisms. Trees may also form a forest which is a suitable habitat for other organisms.

Importance of Ecosystem Engineers

Ecosystem engineers have greater influence and impact on other organisms living in the same environment with them, especially by providing resources to the organisms. Some of the ecosystem engineers have contributed to species richness at the landscape level. The dams built by beavers have ecological effects on other species since they create habitat and control a number of abiotic resources that other animals can use and can also support species not found anywhere. By conserving ecosystem engineers, protection is extended to the overall diversity of a landscape. The biodiversity of an area can also be affected by the ecosystem engineer's ability to increase the complexity of the ecosystem processes which allow for further species greatness. Beavers have the ability to modify a riparian land and expand on wetland habitat leading to an increase in diversity of habitat creating room for more organisms to inhabit the landscape.

Can Humans also be Ecosystem Engineers?

Humans are the most obvious ecosystem engineers. Human activities have contributed to niche construction. The activities such as agriculture, mining, logging and other activities have significantly changed how humans interact with the environment. Some of the human intervention activities such as environmental preservation and upgrading enables an area to be restored to the previous state. Humans also help in the management of evasive species that would otherwise be dangerous to the environment and organisms inhabiting such area.

Productivity

The productivity of an ecosystem refers to the rate of production i.e. the amount of organic matter accumulated in any unit time. Productivity is of following types:

1. Primary Productivity. It is associated with the producers which are autotrophic, most of which are photosynthetic, and to a much lesser extent the chemosynthetic microorganisms. These are the green plants, higher macrophytes as well as lower forms, the phytoplanktons and some photosynthetic bacteria. Primary productivity is defined as "the rate at which radiant energy is stored by photosynthetic and chemosynthetic activities of the producers." Primary productivity is further distinguished as:

(a) Gross Primary Productivity. It is the total rate of photosynthesis including the organic matter used up in respiration during the measurement period. This is also sometimes referred to as Total (Gross) Photosynthesis or Total Assimilation. It depends on the chlorophyll content. It is estimated in terms of either chlorophyll content as, Chl/g dry weight/unit area, or photosynthetic number i.e. amount of CO_2 fixed/g Chl/hour.

(b) Net primary Productivity. It is the rate of storage of organic matter in plant tissues in excess of the respiratory utilization by plants during the measurement period. This is thus the rate of increase of biomass and is also known as Apparent Photosynthesis or Net assimilation. Thus net primary productivity refers to balance between gross photosynthesis and respiration and other plant losses as death etc.

2. Secondary Productivity. It refers to the consumers or heterotrophs. These are the rates of energy storage at consumer level. Since consumers only utilize food material (already produced) in their respiration, simply converting the food matter to different tissues by an overall process, secondary productivity is not divided into 'gross' and 'net' amounts. Thus, some ecologists prefer to use the term assimilation rather than production at this level. Secondary productivity actually remains mobile (i.e. keeps moving from one organism to another) and does not live in situ like the primary productivity.

3. Net Productivity. It refers to the rate of storage of organic matter not used by the heterotrophs (consumers) i.e. equivalent to net primary production minus consumption by the heterotrophs during the unit period, as a season or year etc. It is thus the rate of increase of biomass of the primary producers which has been left over by the consumers. Net productivity is generally expressed as production of C g/m2/day which may then be consolidated on month, season or year basis.

Nutrient Cycle

Nutrient cycling is one of the most important processes that occur in an ecosystem. The nutrient cycle describes the use, movement, and recycling of nutrients in the environment. Valuable elements such as carbon, oxygen, hydrogen, phosphorus, and nitrogen are essential to life and must be recycled in order for organisms to exist. Nutrient cycles are inclusive of both living and non-living components and involve biological, geological, and chemical processes. For this reason, these nutrient circuits are known as biogeochemical cycles.

Biogeochemical Cycles

Biogeochemical cycles can be categorized into two main types: global cycles and local cycles. Elements such as carbon, nitrogen, oxygen, and hydrogen are recycled through abiotic environments including the atmosphere, water, and soil. Since the atmosphere is the main abiotic environment from which these elements are harvested, their cycles are of a global nature. These elements may travel over large distances before they are taken up by biological organisms. The soil is the main abiotic environment for the recycling of elements such as phosphorus, calcium, and potassium. As such, their movement is typically over a local region.

Carbon Cycle

Carbon is essential to all life as it is the main constituent of living organisms. It serves as the backbone component for all organic polymers, including carbohydrates, proteins, and lipids. Carbon compounds, such as carbon dioxide (CO_2) and methane (CH_4), circulate in the atmosphere and influence global climates. Carbon is circulated between living and nonliving components of the ecosystem primarily through the processes of photosynthesis and respiration. Plants and other photosynthetic organisms obtain CO_2 from their environment and use it to build biological materials. Plants, animals, and decomposers (bacteria and fungi) return CO_2 to the atmosphere through respiration. The movement of carbon through biotic components of the environment is known as the fast carbon cycle. It takes considerably less time for carbon to move through the biotic elements of the cycle than it takes for it to move through the abiotic elements. It can take as long as 200 million years for carbon to move through abiotic elements such as rocks, soil, and oceans. Thus, this circulation of carbon is known as the slow carbon cycle.

Carbon Cycles Through the Environment as Follows:

- CO_2 is removed from the atmosphere by photosynthetic organisms (plants, cyanobacteria, etc.) and used to generate organic molecules and build biological mass.

- Animals consume the photosynthetic organisms and acquire the carbon stored within the producers.

- CO_2 is returned to the atmosphere via respiration in all living organisms.

- Decomposers break down dead and decaying organic matter and release CO_2.

- Some CO_2 is returned to the atmosphere via the burning of organic matter (forest fires).

- CO_2 trapped in rock or fossil fuels can be returned to the atmosphere via erosion, volcanic eruptions, or fossil fuel combustion.

Nitrogen Cycle

Similar to carbon, nitrogen is a necessary component of biological molecules. Some of these molecules include amino acids and nucleic acids. Although nitrogen (N_2) is abundant in the atmosphere, most living organisms can not use nitrogen in this form to synthesize organic compounds. Atmospheric nitrogen must first be fixed, or converted to ammonia (NH_3) by certain bacteria.

Nitrogen cycles through the environment as follows:

- Atmospheric nitrogen (N_2) is converted to ammonia (NH_3) by nitrogen-fixing bacteria in aquatic and soil environments. These organisms use nitrogen to synthesise the biological molecules they need to survive.

- NH_3 is subsequently converted to nitrite and nitrate by bacteria known as nitrifying bacteria.

- Plants obtain nitrogen from the soil by absorbing ammonium (NH_4-) and nitrate through their roots. Nitrate and ammonium are used to produce organic compounds.

- Nitrogen in its organic form is obtained by animals when they consume plants or animals.

- Decomposers return NH3 to the soil by decomposing solid waste and dead or decaying matter.

- Nitrifying bacteria convert NH3 to nitrite and nitrate.

- Denitrifying bacteria convert nitrite and nitrate to N2, releasing N2 back into the atmosphere.

Other Chemical Cycles

Oxygen and phosphorus are elements that are also essential to biological organisms. The vast majority of atmospheric oxygen (O_2) is derived from photosynthesis. Plants and other photosynthetic organisms use CO_2, water, and light energy to produce glucose and O_2. Glucose is used to synthesize organic molecules, while O_2 is released into the atmosphere. Oxygen is removed from the atmosphere through decomposition processes and respiration in living organisms.

Phosphorus is a component of biological molecules such as RNA, DNA, phospholipids, and adenosine triphosphate (ATP). ATP is a high energy molecule produced by the processes of cellular respiration and fermentation. In the phosphorus cycle, phosphorus is circulated mainly through soil, rocks, water, and living organisms. Phosphorus is found organically in the form of the phosphate ion (PO_4^{3-}). Phosphorus is added to soil and water by runoff resulting from the weathering of rocks that contain phosphates. PO_4^{3-} is absorbed from the soil by plants and obtained by consumers through the consumption of plants and other animals. Phosphates are added back to the soil through decomposition. Phosphates may also become trapped in sediments in aquatic environments. These phosphate containing sediments form new rocks over time.

Terrestrial Production

An oak tree; a typical modern, terrestrial autotroph

On the land, almost all primary production is now performed by vascular plants, with a small fraction coming from algae and non-vascular plants such as mosses and liverworts. Before the evolution of vascular plants, non-vascular plants likely played a more significant role. Primary production on land is a function of many factors, but principally local hydrology and temperature (the latter covaries to an extent with light, specifically photosynthetically active radiation (PAR), the source of energy for photosynthesis). While plants cover much of the Earth's surface, they are strongly curtailed wherever temperatures are too extreme or where necessary plant resources (principally water and PAR) are limiting, such as deserts or polar regions.

Water is "consumed" in plants by the processes of photosynthesis and transpiration. The latter process (which is responsible for about 90% of water use) is driven by the evaporation of water from the leaves of plants. Transpiration allows plants to transport water and mineral nutrients from the soil to growth regions, and also cools the plant. Diffusion of water vapour out of a leaf, the force that drives transpiration, is regulated by structures known as stomata. These structure also regulate the diffusion of carbon dioxide from the atmosphere into the leaf, such that decreasing water loss (by partially closing stomata) also decreases carbon dioxide gain. Certain plants use alternative forms of photosynthesis, called Crassulacean acid metabolism (CAM) and C4. These employ physiological and anatomical adaptations to increase water-use efficiency and allow increased primary production to take place under conditions that would normally limit carbon fixation by C3 plants (the majority of plant species).

Oceanic Production

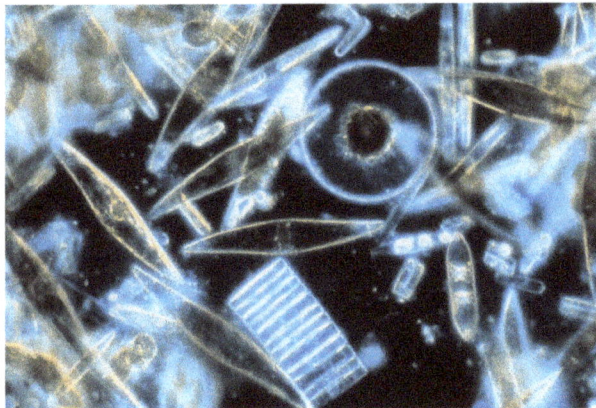

Marine diatoms; an example of planktonic microalgae

In a reversal of the pattern on land, in the oceans, almost all photosynthesis is performed by algae, with a small fraction contributed by vascular plants and other groups. Algae encompass a diverse range of organisms, ranging from single floating cells to attached seaweeds. They include photoautotrophs from a variety of groups. Eubacteria are important photosynthetizers in both oceanic and terrestrial ecosystems, and while some archaea are phototrophic, none are known to utilise oxygen-evolving photosynthesis. A number of eukaryotes are significant contributors to primary production in the ocean, including green algae, brown algae and red algae, and a diverse group of unicellular groups. Vascular plants are also represented in the ocean by groups such as the seagrasses.

Unlike terrestrial ecosystems, the majority of primary production in the ocean is performed by free-living microscopic organisms called phytoplankton. Larger autotrophs, such as the seagrasses and macroalgae (seaweeds) are generally confined to the littoral zone and adjacent shallow waters, where they can attach to the underlying substrate but still be within the photic zone. There are exceptions, such as Sargassum, but the vast majority of free-floating production takes place within microscopic organisms.

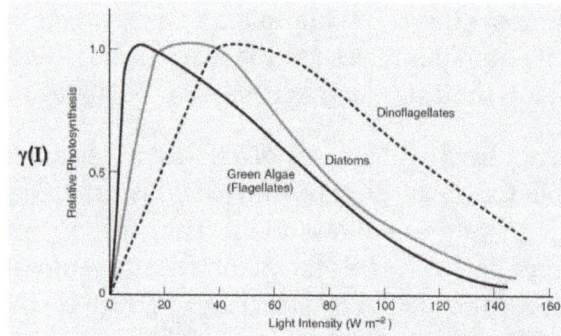

Differences in relative photosynthesis between plankton species under different irradiance

The factors limiting primary production in the ocean are also very different from those on land. The availability of water, obviously, is not an issue (though its salinity can be). Similarly, temperature, while affecting metabolic rates, ranges less widely in the ocean than on land because the heat capacity of seawater buffers temperature changes, and the formation of sea ice insulates it at lower temperatures. However, the availability of light, the source of energy for photosynthesis, and mineral nutrients, the building blocks for new growth, play crucial roles in regulating primary production in the ocean. Available Earth System Models suggest that ongoing ocean bio-geochemical changes could trigger reductions in ocean NPP between 3% and 10% of current values depending on the emissions scenario.

Light

A kelp forest; an example of attached macroalgae

The sunlit zone of the ocean is called the photic zone (or euphotic zone). This is a relatively thin layer (10–100 m) near the ocean's surface where there is sufficient light for photosynthesis to occur. For practical purposes, the thickness of the photic zone is typically defined by the depth at which light reaches 1% of its surface value. Light is attenuated down the water column by its absorption or scattering by the water itself, and by dissolved or particulate material within it (including phytoplankton).

Net photosynthesis in the water column is determined by the interaction between the photic zone and the mixed layer. Turbulent mixing by wind energy at the ocean's surface homogenises the water column vertically until the turbulence dissipates (creating the aforementioned mixed layer). The deeper the mixed layer, the lower the average amount of light intercepted by phytoplankton within it. The mixed layer can vary from being shallower than the photic zone, to being much deeper than the photic zone. When it is much deeper than the photic zone, this results in phytoplankton spending too much time in the dark for net growth to occur. The maximum depth of the mixed layer in which net growth can occur is called the critical depth. As long as there are adequate nutrients available, net primary production occurs whenever the mixed layer is shallower than the critical depth.

Both the magnitude of wind mixing and the availability of light at the ocean's surface are affected across a range of space- and time-scales. The most characteristic of these is the seasonal cycle (caused by the consequences of the Earth's axial tilt), although wind magnitudes additionally have strong spatial components. Consequently, primary production in temperate regions such as the North Atlantic is highly seasonal, varying with both incident light at the water's surface (reduced in winter) and the degree of mixing (increased in winter). In tropical regions, such as the gyres in the middle of the major basins, light may only vary slightly across the year, and mixing may only occur episodically, such as during large storms or hurricanes.

Nutrients

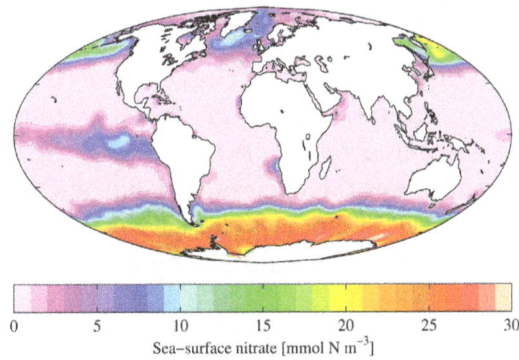

Annual mean sea surface nitrate for the World Ocean

Mixing also plays an important role in the limitation of primary production by nutrients. Inorganic nutrients, such as nitrate, phosphate and silicic acid are necessary for phytoplankton to synthesise their cells and cellular machinery. Because of gravitational sinking of particulate material (such as plankton, dead or fecal material), nutrients are constantly lost from the photic zone, and are only replenished by mixing or upwelling of deeper water. This is exacerbated where summertime solar heating and reduced winds increases vertical stratification and leads to a strong thermocline, since this makes it more difficult for wind mixing to entrain deeper water. Consequently, between mixing events, primary production (and the resulting processes that leads to sinking particulate material) constantly acts to consume nutrients in the mixed layer, and in many regions this leads to nutrient exhaustion and decreased mixed layer production in the summer (even in the presence of abundant light). However, as long as the photic zone is deep enough, primary production may continue below the mixed layer where light-limited growth rates mean that nutrients are often more abundant.

Iron

Another factor relatively recently discovered to play a significant role in oceanic primary production is the micronutrient iron. This is used as a cofactor in enzymes involved in processes such as nitrate reduction and nitrogen fixation. A major source of iron to the oceans is dust from the Earth's deserts, picked up and delivered by the wind as aeolian dust.

In regions of the ocean that are distant from deserts or that are not reached by dust-carrying winds (for example, the Southern and North Pacific oceans), the lack of iron can severely limit the amount of primary production that can occur. These areas are sometimes known as HNLC (High-Nutrient, Low-Chlorophyll) regions, because the scarcity of iron both limits phytoplankton growth and leaves a surplus of other nutrients. Some scientists have suggested introducing iron to these areas as a means of increasing primary productivity and sequestering carbon dioxide from the atmosphere.

Measurement

The methods for measurement of primary production vary depending on whether gross vs net production is the desired measure, and whether terrestrial or aquatic systems are the focus. Gross production is almost always harder to measure than net, because of respiration, which is a continuous and ongoing process that consumes some of the products of primary production (i.e. sugars) before they can be accurately measured. Also, terrestrial ecosystems are generally more difficult because a substantial proportion of total productivity is shunted to below-ground organs and tissues, where it is logistically difficult to measure. Shallow water aquatic systems can also face this problem.

Scale also greatly affects measurement techniques. The rate of carbon assimilation in plant tissues, organs, whole plants, or plankton samples can be quantified by biochemically based techniques, but these techniques are decidedly inappropriate for large scale terrestrial field situations. There, net primary production is almost always the desired variable, and estimation techniques involve various methods of estimating dry-weight biomass changes over time. Biomass estimates are often converted to an energy measure, such as kilocalories, by an empirically determined conversion factor.

Terrestrial

In terrestrial ecosystems, researchers generally measure net primary production (NPP). Although its definition is straightforward, field measurements used to estimate productivity vary according to investigator and biome. Field estimates rarely account for below ground productivity, herbivory, turnover, litterfall, volatile organic compounds, root exudates, and allocation to symbiotic microorganisms. Biomass based NPP estimates result in underestimation of NPP due to incomplete accounting of these components. However, many field measurements correlate well to NPP. There are a number of comprehensive reviews of the field methods used to estimate NPP. Estimates of ecosystem respiration, the total carbon dioxide produced by the ecosystem, can also be made with gas flux measurements.

The major unaccounted pool is belowground productivity, especially production and turnover of roots. Belowground components of NPP are difficult to measure. BNPP (below-ground NPP) is often estimated based on a ratio of ANPP:BNPP (above-ground NPP:below-ground NPP) rather than direct measurements.

Gross primary production can be estimated from measurements of net ecosystem exchange (NEE) of carbon dioxide made by the eddy covariance technique. During night, this technique measures all components of ecosystem respiration. This respiration is scaled to day-time values and further subtracted from NEE.

Grasslands

The Konza tallgrass prairie in the Flint Hills of northeastern Kansas

Most frequently, peak standing biomass is assumed to measure NPP. In systems with persistent standing litter, live biomass is commonly reported. Measures of peak biomass are more reliable if the system is predominantly annuals. However, perennial measurements could be reliable if there were a synchronous phenology driven by a strong seasonal climate. These methods may underestimate ANPP in grasslands by as much as 2 (temperate) to 4 (tropical) fold. Repeated measures of standing live and dead biomass provide more accurate estimates of all grasslands, particularly those with large turnover, rapid decomposition, and interspecific variation in timing of peak biomass. Wetland productivity (marshes and fens) is similarly measured. In Europe, annual mowing makes the annual biomass increment of wetlands evident.

Forests

Methods used to measure forest productivity are more diverse than those of grasslands. Biomass increment based on stand specific allometry plus litterfall is considered a suitable although incomplete accounting of above-ground net primary production (ANPP). Field measurements used as a proxy for ANPP include annual litterfall, diameter or basal area increment (DBH or BAI), and volume increment.

Aquatic

In aquatic systems, primary production is typically measured using one of six main techniques:

1. variations in oxygen concentration within a sealed bottle (developed by Gaarder and Gran in 1927)

2. incorporation of inorganic carbon-14 (14C in the form of sodium bicarbonate) into organic matter

3. Stable isotopes of Oxygen (16O, 18O and 17O)

4. fluorescence kinetics (technique still a research topic)

5. Stable isotopes of Carbon (12C and 13C)

6. Oxygen/Argon Ratios

The technique developed by Gaarder and Gran uses variations in the concentration of oxygen under different experimental conditions to infer gross primary production. Typically, three identical transparent vessels are filled with sample water and stoppered. The first is analysed immediately and used to determine the initial oxygen concentration; usually this is done by performing a Winkler titration. The other two vessels are incubated, one each in under light and darkened. After a fixed period of time, the experiment ends, and the oxygen concentration in both vessels is measured. As photosynthesis has not taken place in the dark vessel, it provides a measure of ecosystem respiration. The light vessel permits both photosynthesis and respiration, so provides a measure of net photosynthesis (i.e. oxygen production via photosynthesis subtract oxygen consumption by respiration). Gross primary production is then obtained by adding oxygen consumption in the dark vessel to net oxygen production in the light vessel.

The technique of using ^{14}C incorporation (added as labelled Na_2CO_3) to infer primary production is most commonly used today because it is sensitive, and can be used in all ocean environments. As ^{14}C is radioactive (via beta decay), it is relatively straightforward to measure its incorporation in organic material using devices such as scintillation counters.

Depending upon the incubation time chosen, net or gross primary production can be estimated. Gross primary production is best estimated using relatively short incubation times (1 hour or less), since the loss of incorporated ^{14}C (by respiration and organic material excretion / exudation) will be more limited. Net primary production is the fraction of gross production remaining after these loss processes have consumed some of the fixed carbon.

Loss processes can range between 10-60% of incorporated ^{14}C according to the incubation period, ambient environmental conditions (especially temperature) and the experimental species used. Aside from those caused by the physiology of the experimental subject itself, potential losses due to the activity of consumers also need to be considered. This is particularly true in experiments making use of natural assemblages of microscopic autotrophs, where it is not possible to isolate them from their consumers.

The methods based on stable isotopes and O_2/Ar ratios have the advantage of providing estimates of respiration rates in the light without the need of incubations in the dark. Among them, the method of the triple oxygen isotopes and O_2/Ar have the additional advantage of not needing incubations in closed containers and O_2/Ar can even be measured continuously at sea using equilibrator inlet mass spectrometry (EIMS) or a membrane inlet mass spectrometry (MIMS). However, if results relevant to the carbon cycle are desired, it is probably better to rely on methods based on carbon (and not oxygen) isotopes. It is important to notice that the method based on carbon stable isotopes is not simply an adaptation of the classic ^{14}C method, but an entirely different approach that does not suffer from the problem of lack of account of carbon recycling during photosynthesis.

Global

As primary production in the biosphere is an important part of the carbon cycle, estimating it at the global scale is important in Earth system science. However, quantifying primary production at this scale is difficult because of the range of habitats on Earth, and because of the impact of weather events (availability of sunlight, water) on its variability. Using satellite-derived estimates of the Normalized Difference Vegetation Index (NDVI) for terrestrial habitats and sea-surface chlorophyll for the oceans, it is estimated that the total (photoautotrophic) primary production for the Earth was 104.9 Gt C yr⁻¹. Of this, 56.4 Gt C yr⁻¹ (53.8%), was the product of terrestrial organisms, while the remaining 48.5 Gt C yr⁻¹, was accounted for by oceanic production.

Scaling ecosystem-level GPP estimations based on eddy covariance measurements of net ecosystem exchange to regional and global values using spatial details of different predictor variables, such as climate variables and remotely sensed fAPAR or LAI led to a terrestrial gross primary production of 123±8 Gt carbon (NOT carbon dioxide) per year during 1998-2005.

In areal terms, it was estimated that land production was approximately 426 g C m⁻² yr⁻¹ (excluding areas with permanent ice cover), while that for the oceans was 140 g C m⁻² yr⁻¹. Another significant difference between the land and the oceans lies in their standing stocks - while accounting for almost half of total production, oceanic autotrophs only account for about 0.2% of the total biomass.

Estimates

Primary productivity can be estimated by a variety of proxies. One that has particular relevance to the geological record is Barium, whose concentration in marine sediments rises in line with primary productivity at the surface.

Primary Production and Plant Biomass for the Earth

Ecosystem type	Area (10^6 km²)	Mean NPP (g/m²/yr)	World NPP (10^9 tons/yr)	Mean biomass (kg/m²)	World biomass (10^9 tons)
Tropical rainforest	17.0	2,200	37.4	45	763
Tropical seasonal forest	7.5	1,600	12.0	35	260
Temperate evergreen forest	5.0	1,300	6.5	35	175
Temperate deciduous forest	7.0	1,200	8.4	30	210
Boreal forest	12.0	800	9.6	20	240
Woodlands and shrublands	8.5	700	6.0	6	50
Savanna	15.0	900	13.6	4	60
Temperate grasslands	9.0	600	5.4	1.6	14
Tundra and alpine	8.0	140	1.1	0.6	5
Desert and semi-desert	18.0	90	1.6	0.7	13
Extreme desert and ice	24.0	3	0.07	0.02	0.5
Cultivated land	14.0	650	9.1	1.0	14
Swamp and wetland	2.0	2,000	4.0	12.3	30
Lakes and streams	2.0	250	0.5	0.02	0.05
Total Continental	149	773	115	12.3	1837
Open ocean	332.0	125	41.5	0.003	1.0
Upwelling zones	0.4	500	0.2	0.02	0.008
Continental shelf	26.6	360	9.6	0.01	0.27
Algal bed and reef	0.6	2,500	1.6	2.0	1.2
Estuaries	1.4	1,500	2.1	1.0	1.4

| Total marine | 361 | 152 | 55.0 | 0.01 | 3.9 |
| Grand total | 510 | 333 | 170 | 3.6 | 1841 |

From R.H. Whittaker, quoted in Peter Stiling (1996), "Ecology: Theories and Applications" (Prentice Hall).

Primary Producers

Primary producers form the bases of food webs and entire ecosystems. They produce their own food either via photosynthesis or chemosynthesis, depending on the availability of sunlight. Primary producers provide food for primary consumers or herbivores, which in turn provide food for carnivores, consumers at secondary or tertiary trophic levels.

The Role of Primary Producers

Primary producers form the basis of a food web or tropic levels. Whether it's green plants using photosynthesis or benthic bacterial mats, these organisms provide food for consumers. Primary consumers are herbivore animals that eat primary producers. Carnivorous secondary consumers eat primary consumers. Tertiary consumers eat the secondary consumers. Eventually, the top member of a food web will die, and its body will feed decomposers. These decomposers provide organic material, converted for use by the next cohort of primary consumers, and begin the cycle again. Energy passes through each trophic level. However, throughout each step of the food web, energy is lost either as heat, for life processes such as movement, or in feces and decaying remains.

Producers using Photosynthesis

Plants, algae and photosynthetic bacteria comprise primary producers that use the sun's energy for photosynthesis. Both terrestrial and aquatic primary producers perform this activity. In the water column, phytoplankton living closer to the water's surface access the sun's energy. Other primary producers in aquatic environments include seaweeds, algae, green plants, and blue-green or purple bacteria. They photosynthesize carbohydrates needed for energy. Phytoplankton comprise the foundation of ocean ecosystems, as many creatures from zooplankton to jellies to whales rely on them for food.

Producers using Chemosynthesis

Some primary producers, such as some species of bacteria, do not use the sun for energy nor for fixing carbon dioxide. This process, chemosynthesis, occurs in the absence of sunlight, such as deep in the ocean around hydrothermal vents or cold water methane seeps. These primary producers metabolize inorganic material such as nitrogen, sulfur or iron to make food from carbon dioxide. These types of producers are called chemoautotrophs, and they form the basis of food webs for octopi, crabs and other sea animals. Chemoautotrophs can also be found in other hydrogen sulfide-rich ecosystems such as salt marshes.

Interconnected Success

The quality of primary producers directly affects the success of secondary consumers, and as a result, the entire food chain. Because of the interconnected nature of food webs, it is crucial for primary producers to remain intact. Even some of the smallest primary producers, such as phytoplankton, possess far-reaching effects. Many aquatic food webs rely on phytoplankton as primary producers. The different biochemical makeup of phytoplankton affects their nutritional value for zooplankton that feed on them. Biodiversity of primary producers, whether aquatic or terrestrial, yields greater success for herbivores because the animals have more nutritious options available.

Primary producers also rely on the work of decomposers to survive. Decomposers fix nitrogen so that it is available for primary producers. Various organisms work the soil to decompose organic matter and enhance its nutrients so that plants can grow and provide the basis of a food web. While different organisms can take the place of those that grow scarce, maintaining and protecting primary producers yields sustainable ecosystems.

Food Web

Food web is an important ecological concept. Basically, food web represents feeding relationships within a community (Smith and Smith 2009). It also implies the transfer of food energy from its source in plants through herbivores to carnivores (Krebs 2009). Normally, food webs consist of a number of food chains meshed together. Each food chain is a descriptive diagram including a series of arrows, each pointing from one species to another, representing the flow of food energy from one feeding group of organisms to another.

There are two types of food chains: the grazing food chain, beginning with autotrophs, and the detrital food chain, beginning with dead organic matter (Smith & Smith 2009). In a grazing food chain, energy and nutrients move from plants to the herbivores consuming them, and to the carnivores or omnivores preying upon the herbivores. In a detrital food chain, dead organic matter of plants and animals is broken down by decomposers, e.g., bacteria and fungi, and moves to detritivores and then carnivores.

Food web offers an important tool for investigating the ecological interactions that define energy flows and predator-prey relationship (Cain et al. 2008). Figure shows a simplified food web in a desert ecosystem. In this food web, grasshoppers feed on plants; scorpions prey on grasshoppers; kit foxes prey on scorpions. While the food web showed here is a simple one, most feed webs are complex and involve many species with both strong and weak interactions among them (Pimm et al. 1991). For example, the predators of a scorpion in a desert ecosystem might be a golden eagle, an owl, a roadrunner, or a fox.

The idea to apply the food chains to ecology and to analyze its consequences was first proposed by Charles Elton (Krebs 2009). In 1927, he recognized that the length of these food chains was mostly limited to 4 or 5 links and the food chains were not isolated, but hooked together into food webs (which he called "food cycles").

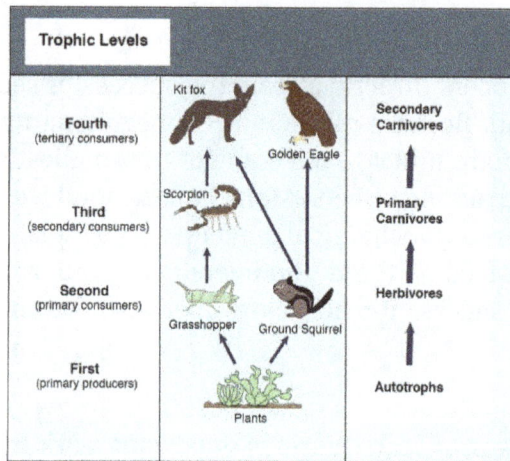

Figure: A simple six-member food web for a representative desert grassland.

Taxonomy of a Food Web

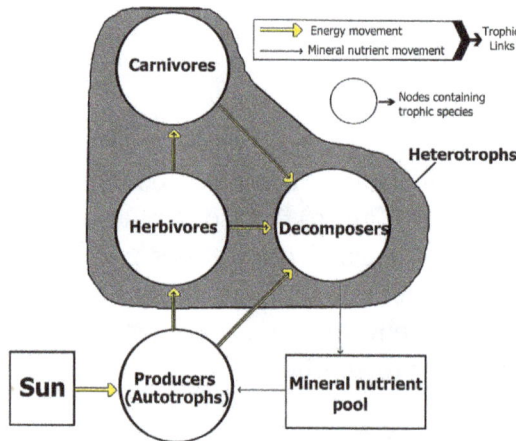

A simplified food web illustrating a three trophic food chain
(producers-herbivores-carnivores) linked to decomposers.

The movement of mineral nutrients is cyclic, whereas the movement of energy is unidirectional and noncyclic. Trophic species are encircled as nodes and arrows depict the links.

Links in food webs map the feeding connections (who eats whom) in an ecological community. Food cycle is an obsolete term that is synonymous with food web. Ecologists can broadly group all life forms into one of two trophic layers, the autotrophs and the heterotrophs. Autotrophs produce more biomass energy, either chemically without the sun's energy or by capturing the sun's energy in photosynthesis, than they use during metabolic respiration. Heterotrophs consume rather than produce biomass energy as they metabolize, grow, and add to levels of secondary production. A food web depicts a collection of polyphagous heterotrophic consumers that network and cycle the flow of energy and nutrients from a productive base of self-feeding autotrophs.

The base or basal species in a food web are those species without prey and can include autotrophs or saprophytic detritivores (i.e., the community of decomposers in soil, biofilms, and periphyton). Feeding connections in the web are called trophic links. The number of trophic links per consumer is a measure of food web connectance. Food chains are nested within the trophic links of food webs. Food chains are linear (noncyclic) feeding pathways that trace monophagous consumers from a base species up to the top consumer, which is usually a larger predatory carnivore.

Linkages connect to nodes in a food web, which are aggregates of biological taxa called trophic species. Trophic species are functional groups that have the same predators and prey in a food web. Common examples of an aggregated node in a food web might include parasites, microbes, decomposers, saprotrophs, consumers, or predators, each containing many species in a web that can otherwise be connected to other trophic species.

Trophic Levels

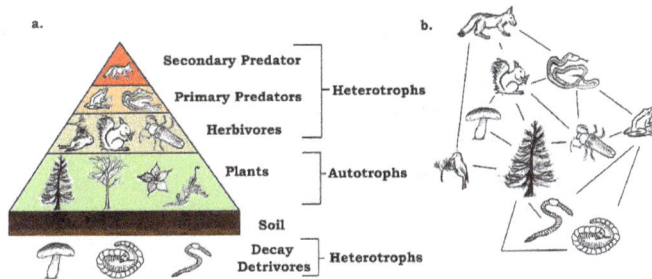

A trophic pyramid (a) and a simplified community food web (b) illustrating ecological relations among creatures that are typical of a northern Boreal terrestrial ecosystem.

The trophic pyramid roughly represents the biomass (usually measured as total dry-weight) at each level. Plants generally have the greatest biomass. Names of trophic categories are shown to the right of the pyramid. Some ecosystems, such as many wetlands, do not organize as a strict pyramid, because aquatic plants are not as productive as long-lived terrestrial plants such as trees. Ecological trophic pyramids are typically one of three kinds: 1) pyramid of numbers, 2) pyramid of biomass, or 3) pyramid of energy.

Food webs have trophic levels and positions. Basal species, such as plants, form the first level and are the resource limited species that feed on no other living creature in the web. Basal species can be autotrophs or detritivores, including "decomposing organic material and its associated microorganisms which we defined as detritus, micro-inorganic material and associated microorganisms (MIP), and vascular plant material." Most autotrophs capture the sun's energy in chlorophyll, but some autotrophs (the chemolithotrophs) obtain energy by the chemical oxidation of inorganic compounds and can grow in dark environments, such as the sulfur bacterium Thiobacillus, which lives in hot sulfur springs. The top level has top (or apex) predators which no other species kills directly for its food resource needs. The intermediate levels are filled with omnivores that feed on more than one trophic level and cause energy to flow through a number of food pathways starting from a basal species.

In the simplest scheme, the first trophic level (level 1) is plants, then herbivores (level 2), and then carnivores (level 3). The trophic level is equal to one more than the chain length, which is the number of links connecting to the base. The base of the food chain (primary producers or detritivores) is set at zero. Ecologists identify feeding relations and organize species into trophic species through extensive gut content analysis of different species. The technique has been improved through the use of stable isotopes to better trace energy flow through the web. It was once thought that omnivory was rare, but recent evidence suggests otherwise. This realization has made trophic classifications more complex.

Trophic Dynamics

The trophic level concept was introduced in a historical landmark paper on trophic dynamics in 1942 by Raymond L. Lindeman. The basis of trophic dynamics is the transfer of energy from one part of the ecosystem to another. The trophic dynamic concept has served as a useful quantitative heuristic, but it has several major limitations including the precision by which an organism can be allocated to a specific trophic level. Omnivores, for example, are not restricted to any single level. Nonetheless, recent research has found that discrete trophic levels do exist, but "above the herbivore trophic level, food webs are better characterized as a tangled web of omnivores."

A central question in the trophic dynamic literature is the nature of control and regulation over resources and production. Ecologists use simplified one trophic position food chain models (producer, carnivore, decomposer). Using these models, ecologists have tested various types of ecological control mechanisms. For example, herbivores generally have an abundance of vegetative resources, which meant that their populations were largely controlled or regulated by predators. This is known as the top-down hypothesis or 'green-world' hypothesis. Alternatively to the top-down hypothesis, not all plant material is edible and the nutritional quality or antiherbivore defenses of plants (structural and chemical) suggests a bottom-up form of regulation or control. Recent studies have concluded that both "top-down" and "bottom-up" forces can influence community structure and the strength of the influence is environmentally context dependent. These complex multitrophic interactions involve more than two trophic levels in a food web.

Another example of a multi-trophic interaction is a trophic cascade, in which predators help to increase plant growth and prevent overgrazing by suppressing herbivores. Links in a food-web illustrate direct trophic relations among species, but there are also indirect effects that can alter the abundance, distribution, or biomass in the trophic levels. For example, predators eating herbivores indirectly influence the control and regulation of primary production in plants. Although the predators do not eat the plants directly, they regulate the population of herbivores that are directly linked to plant trophism. The net effect of direct and indirect relations is called trophic cascades. Trophic cascades are separated into species-level cascades, where only a subset of the food-web dynamic is impacted by a change in population numbers, and community-level cascades, where a change in population numbers has a dramatic effect on the entire food-web, such as the distribution of plant biomass.

Energy Flow and Biomass

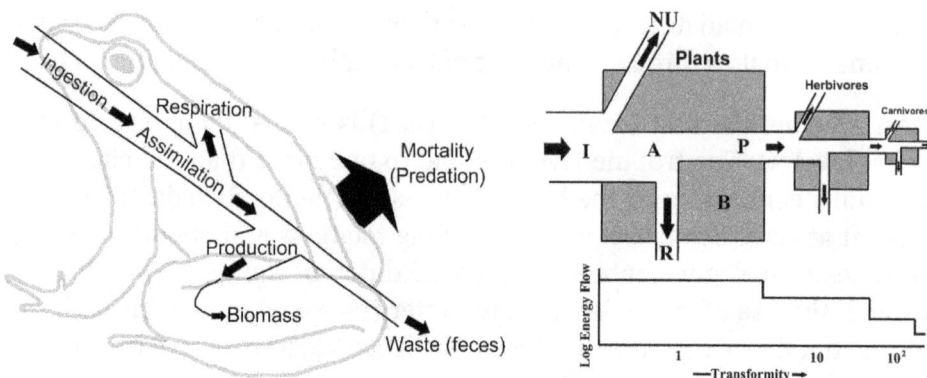

Left: Energy flow diagram of a frog. The frog represents a node in an extended food web. The energy ingested is utilized for metabolic processes and transformed into biomass. The energy flow continues on its path if the frog is ingested by predators, parasites, or as a decaying carcass in soil. This energy flow diagram illustrates how energy is lost as it fuels the metabolic process that transform the energy and nutrients into biomass.

Right: An expanded three link energy food chain (1. plants, 2. herbivores, 3. carnivores) illustrating the relationship between food flow diagrams and energy transformity. The transformity of energy becomes degraded, dispersed, and diminished from higher quality to lesser quantity as the energy within a food chain flows from one trophic species into another. Abbreviations: I=input, A=assimilation, R=respiration, NU=not utilized, P=production, B=biomass.

Food webs depict energy flow via trophic linkages. Energy flow is directional, which contrasts against the cyclic flows of material through the food web systems. Energy flow "typically includes production, consumption, assimilation, non-assimilation losses (feces), and respiration (maintenance costs)." In a very general sense, energy flow (E) can be defined as the sum of metabolic production (P) and respiration (R), such that E=P+R.

Biomass represents stored energy. However, concentration and quality of nutrients and energy is variable. Many plant fibers, for example, are indigestible to many herbivores leaving grazer community food webs more nutrient limited than detrital food webs where bacteria are able to access and release the nutrient and energy stores. "Organisms usually extract energy in the form of carbohydrates, lipids, and proteins. These polymers have a dual role as supplies of energy as well as building blocks; the part that functions as energy supply results in the production of nutrients (and carbon dioxide, water, and heat). Excretion of nutrients is, therefore, basic to metabolism." The units in energy flow webs are typically a measure mass or energy per m² per unit time. Different consumers are going to have different metabolic assimilation efficiencies in their diets. Each trophic level transforms energy into biomass. Energy flow diagrams illustrate the rates and efficiency of transfer from one trophic level into another and up through the hierarchy.

It is the case that the biomass of each trophic level decreases from the base of the chain to the top. This is because energy is lost to the environment with each transfer as entropy increases. About eighty to ninety percent of the energy is expended for the organism's life processes or is lost as heat or waste. Only about ten to twenty percent of the organism's energy is generally passed to the next organism. The amount can be less than one percent in animals consuming less digestible plants, and it can be as high as forty percent in zooplankton consuming phytoplankton. Graphic representations of the biomass or productivity at each tropic level are called ecological pyramids or trophic pyramids. The transfer of energy from primary producers to top consumers can also be characterized by energy flow diagrams.

Food Chain

A common metric used to quantify food web trophic structure is food chain length. Food chain length is another way of describing food webs as a measure of the number of species encountered as energy or nutrients move from the plants to top predators. There are different ways of calculating food chain length depending on what parameters of the food web dynamic are being considered: connectance, energy, or interaction. In its simplest form, the length of a chain is the number of links between a trophic consumer and the base of the web. The mean chain length of an entire web is the arithmetic average of the lengths of all chains in a food web.

In a simple predator-prey example, a deer is one step removed from the plants it eats (chain length = 1) and a wolf that eats the deer is two steps removed from the plants (chain length = 2). The relative amount or strength of influence that these parameters have on the food web address questions about:

- the identity or existence of a few dominant species (called strong interactors or keystone species)

- the total number of species and food-chain length (including many weak interactors) and

- how community structure, function and stability is determined.

Ecological Pyramids

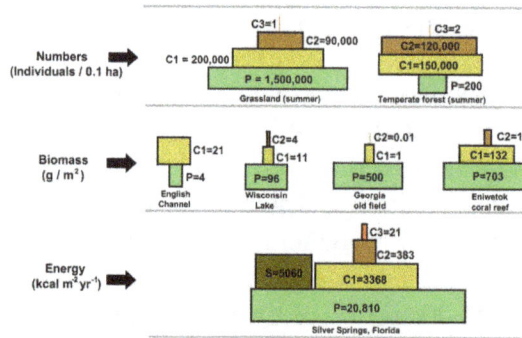

Top Left: A four level trophic pyramid sitting on a layer of soil and its community of decomposers. Top right: A three layer trophic pyramid linked to the biomass and energy flow concepts. Bottom: Illustration of a range of ecological pyramids, including top pyramid of numbers, middle pyramid of biomass, and bottom pyramid of energy. The terrestrial forest (summer) and the English Channel ecosystems exhibit inverted pyramids.Note: trophic levels are not drawn to scale and the pyramid of numbers excludes microorganisms and soil animals. Abbreviations: P=Producers, C1=Primary consumers, C2=Secondary consumers, C3=Tertiary consumers, S=Saprotrophs.

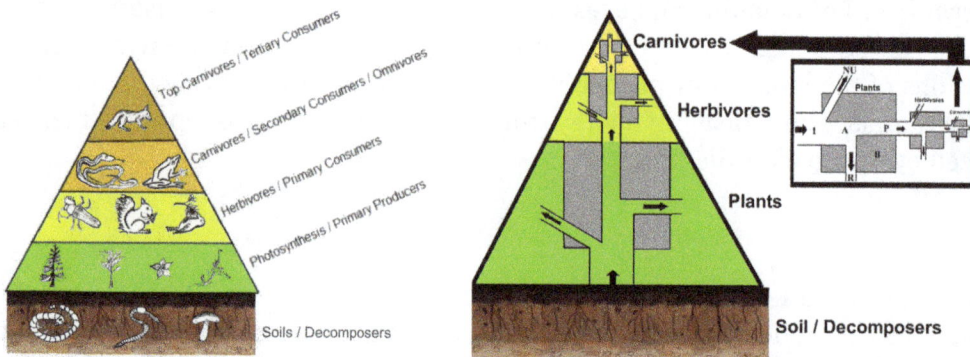

In a pyramid of numbers, the number of consumers at each level decreases significantly, so that a single top consumer, (e.g., a polar bear or a human), will be supported by a much larger number of separate producers. There is usually a maximum of four or five links in a food chain, although food chains in aquatic ecosystems are more often longer than those on land. Eventually, all the energy in a food chain is dispersed as heat.

Ecological pyramids place the primary producers at the base. They can depict different numerical properties of ecosystems, including numbers of individuals per unit of area, biomass (g/m²), and energy (k cal m⁻² yr⁻¹). The emergent pyramidal arrangement of trophic levels with amounts of energy transfer decreasing as species become further removed from the source of production is one of several patterns that is repeated amongst the planets ecosystems.\ The size of each level in the pyramid generally represents biomass, which can be measured as the dry weight of an organism. Autotrophs may have the highest global proportion of biomass, but they are closely rivaled or surpassed by microbes.

Pyramid structure can vary across ecosystems and across time. In some instances biomass pyramids can be inverted. This pattern is often identified in aquatic and coral reef ecosystems. The pattern of biomass inversion is attributed to different sizes of producers. Aquatic communities are often dominated by producers that are smaller than the consumers that have high growth rates. Aquatic producers, such as planktonic algae or aquatic plants, lack the large accumulation of secondary growth as exists in the woody trees of terrestrial ecosystems. However, they are able to reproduce quickly enough to support a larger biomass of grazers. This inverts the pyramid. Primary consumers have longer lifespans and slower growth rates that accumulates more biomass than the producers they consume. Phytoplankton live just a few days, whereas the zooplankton eating the phytoplankton live for several weeks and the fish eating the zooplankton live for several consecutive years. Aquatic predators also tend to have a lower death rate than the smaller consumers, which contributes to the inverted pyramidal pattern. Population structure, migration rates, and environmental refuge for prey are other possible causes for pyramids with biomass inverted. Energy pyramids, however, will always have an upright pyramid shape if all sources of food energy are included and this is dictated by the second law of thermodynamics.

Material Flux and Recycling

Many of the Earth's elements and minerals (or mineral nutrients) are contained within the tissues and diets of organisms. Hence, mineral and nutrient cycles trace food web energy pathways. Ecologists employ stoichiometry to analyze the ratios of the main elements found in all organisms: carbon (C), nitrogen (N), phosphorus (P). There is a large transitional difference between many terrestrial and aquatic systems as C:P and C:N ratios are much higher in terrestrial systems while N:P ratios are equal between the two systems. Mineral nutrients are the material resources that organisms need for growth, development, and vitality. Food webs depict the pathways of mineral nutrient cycling as they flow through organisms. Most of the primary production in an ecosystem is not consumed, but is recycled by detritus back into useful nutrients. Many of the Earth's microorganisms are involved in the formation of minerals in a process called biomineralization. Bacteria that live in detrital sediments create and cycle nutrients and biominerals. Food web models and nutrient cycles have traditionally been treated separately, but there is a strong functional connection between the two in terms of stability, flux, sources, sinks, and recycling of mineral nutrients.

Kinds of Food Webs

Food webs are necessarily aggregated and only illustrate a tiny portion of the complexity of real ecosystems. For example, the number of species on the planet are likely in the general order of

10[7], over 95% of these species consist of microbes and invertebrates, and relatively few have been named or classified by taxonomists. It is explicitly understood that natural systems are 'sloppy' and that food web trophic positions simplify the complexity of real systems that sometimes over-emphasize many rare interactions. Most studies focus on the larger influences where the bulk of energy transfer occurs. "These omissions and problems are causes for concern, but on present evidence do not present insurmountable difficulties."

Chengjiang Shale   Burgess Shale

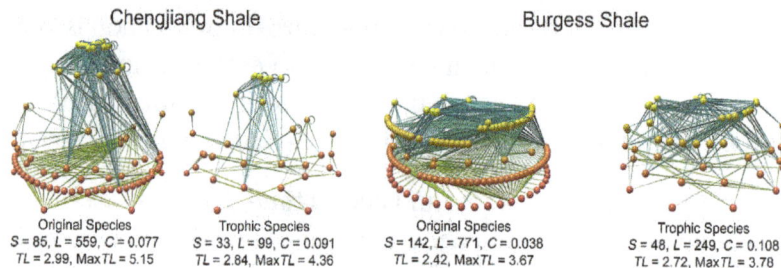

Original Species	Trophic Species	Original Species	Trophic Species
$S = 85, L = 559, C = 0.077$	$S = 33, L = 99, C = 0.091$	$S = 142, L = 771, C = 0.038$	$S = 48, L = 249, C = 0.108$
$TL = 2.99, MaxTL = 5.15$	$TL = 2.84, MaxTL = 4.36$	$TL = 2.42, MaxTL = 3.67$	$TL = 2.72, MaxTL = 3.78$

Paleoecological studies can reconstruct fossil food-webs and trophic levels. Primary producers form the base (red spheres), predators at top (yellow spheres), the lines represent feeding links. Original food-webs (left) are simplified (right) by aggregating groups feeding on common prey into coarser grained trophic species.

There are Different Kinds or Categories of Food Webs:

- Source web - one or more node(s), all of their predators, all the food these predators eat, and so on.

- Sink web - one or more node(s), all of their prey, all the food that these prey eat, and so on.

- Community (or connectedness) web - a group of nodes and all the connections of who eats whom.

- Energy flow web - quantified fluxes of energy between nodes along links between a resource and a consumer.

- Paleoecological web - a web that reconstructs ecosystems from the fossil record.

- Functional web - emphasizes the functional significance of certain connections having strong interaction strength and greater bearing on community organization, more so than energy flow pathways. Functional webs have compartments, which are sub-groups in the larger network where there are different densities and strengths of interaction. Functional webs emphasize that "the importance of each population in maintaining the integrity of a community is reflected in its influence on the growth rates of other populations."

Within these categories, food webs can be further organized according to the different kinds of ecosystems being investigated. For example, human food webs, agricultural food webs, detrital food webs, marine food webs, aquatic food webs, soil food webs, Arctic (or polar) food webs, terrestrial food webs, and microbial food webs. These characterizations stem from the ecosystem concept, which assumes that the phenomena under investigation (interactions and feedback loops) are sufficient to explain patterns within boundaries, such as the edge of a forest, an island, a shoreline, or some other pronounced physical characteristic.

Detrital Web

In a detrital web, plant and animal matter is broken down by decomposers, e.g., bacteria and fungi, and moves to detritivores and then carnivores. There are often relationships between the detrital web and the grazing web. Mushrooms produced by decomposers in the detrital web become a food source for deer, squirrels, and mice in the grazing web. Earthworms eaten by robins are detritivores consuming decaying leaves.

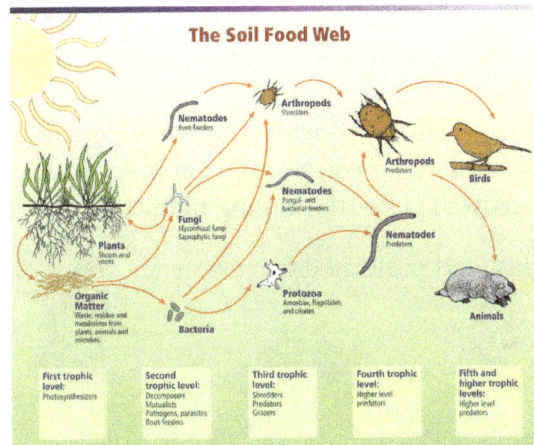

Relationships between soil food web, plants, organic matter, and birds and mammals
Image courtesy of USDA Natural Resources Conservation Service
http://soils.usda.gov/sqi/soil_quality/soil_biology/soil_food_web.html.

An illustration of a soil food web.

"Detritus can be broadly defined as any form of non-living organic matter, including different types of plant tissue (e.g. leaf litter, dead wood, aquatic macrophytes, algae), animal tissue (carrion), dead microbes, faeces (manure, dung, faecal pellets, guano, frass), as well as products secreted, excreted or exuded from organisms (e.g. extra-cellular polymers, nectar, root exudates and leachates, dissolved organic matter, extra-cellular matrix, mucilage). The relative importance of these forms of detritus, in terms of origin, size and chemical composition, varies across ecosystems."

Quantitative Food Webs

Ecologists collect data on trophic levels and food webs to statistically model and mathematically calculate parameters, such as those used in other kinds of network analysis (e.g., graph theory), to study emergent patterns and properties shared among ecosystems. There are different ecological dimensions that can be mapped to create more complicated food webs, including: species composition (type of species), richness (number of species), biomass (the dry weight of plants and animals), productivity (rates of conversion of energy and nutrients into growth), and stability (food webs over time). A food web diagram illustrating species composition shows how change in a single species can directly and indirectly influence many others. Microcosm studies are used to simplify food web research into semi-isolated units such as small springs, decaying logs, and laboratory experiments using organisms that reproduce quickly, such as daphnia feeding on algae grown under controlled environments in jars of water.

While the complexity of real food webs connections are difficult to decipher, ecologists have found mathematical models on networks an invaluable tool for gaining insight into the structure, stability, and laws of food web behaviours relative to observable outcomes. "Food web theory centers around the idea of connectance." Quantitative formulas simplify the complexity of food web structure. The number of trophic links (t_L), for example, is converted into a connectance value:

$$C = \frac{t_L}{S(S-1)/2},$$

where, S(S-1)/2 is the maximum number of binary connections among S species. "Connectance (C) is the fraction of all possible links that are realized (L/S^2) and represents a standard measure of food web complexity..." The distance (d) between every species pair in a web is averaged to compute the mean distance between all nodes in a web (D) and multiplied by the total number of links (L) to obtain link-density (LD), which is influenced by scale dependent variables such as species richness. These formulas are the basis for comparing and investigating the nature of non-random patterns in the structure of food web networks among many different types of ecosystems.

Scaling laws, complexity, choas, and patterned correlates are common features attributed to food web structure.

Complexity and Stability

Food webs are complex. Complexity is a measure of an increasing number of permutations and it is also a metaphorical term that conveys the mental intractability or limits concerning unlimited algorithmic possibilities. In food web terminology, complexity is a product of the number of species and connectance. Connectance is "the fraction of all possible links that are realized in a network".These concepts were derived and stimulated through the suggestion that complexity leads to stability in food webs, such as increasing the number of trophic levels in more species rich ecosystems. This hypothesis was challenged through mathematical models suggesting otherwise, but subsequent studies have shown that the premise holds in real systems.

At different levels in the hierarchy of life, such as the stability of a food web, "the same overall structure is maintained in spite of an ongoing flow and change of components." The farther a living system (e.g., ecosystem) sways from equilibrium, the greater its complexity. Complexity has multiple meanings in the life sciences and in the public sphere that confuse its application as a precise term for analytical purposes in science. Complexity in the life sciences (or biocomplexity) is defined by the "properties emerging from the interplay of behavioral, biological, physical, and social interactions that affect, sustain, or are modified by living organisms, including humans".

Several concepts have emerged from the study of complexity in food webs. Complexity explains many principals pertaining to self-organization, non-linearity, interaction, cybernetic feedback, discontinuity, emergence, and stability in food webs. Nestedness, for example, is defined as "a pattern of interaction in which specialists interact with species that form perfect subsets of the species with which generalists interact", "—that is, the diet of the most specialized species is a subset of the diet of the next more generalized species, and its diet a subset of the next more generalized, and so on." Until recently, it was thought that food webs had little nested structure, but empirical evidence shows that many published webs have nested subwebs in their assembly.

Food webs are complex networks. As networks, they exhibit similar structural properties and mathematical laws that have been used to describe other complex systems, such as small world and scale free properties. The small world attribute refers to the many loosely connected nodes, non-random dense clustering of a few nodes (i.e., trophic or keystone species in ecology), and small path length compared to a regular lattice. "Ecological networks, especially mutualistic networks, are generally very heterogeneous, consisting of areas with sparse links among species and distinct areas of tightly linked species. These regions of high link density are often referred to as cliques, hubs, compartments, cohesive sub-groups, or modules...Within food webs, especially in aquatic systems, nestedness appears to be related to body size because the diets of smaller predators tend to be nested subsets of those of larger predators (Woodward & Warren 2007; YvonDurocher et al. 2008), and phylogenetic constraints, whereby related taxa are nested based on their common evolutionary history, are also evident (Cattin et al. 2004)." "Compartments in food webs are subgroups of taxa in which many strong interactions occur within the subgroups and few weak interactions occur between the subgroups. Theoretically, compartments increase the stability in networks, such as food webs."

Food webs are also complex in the way that they change in scale, seasonally, and geographically. The components of food webs, including organisms and mineral nutrients, cross the thresholds of ecosystem boundaries. This has led to the concept or area of study known as cross-boundary subsidy. "This leads to anomalies, such as food web calculations determining that an ecosystem can support one half of a top carnivore, without specifying which end." Nonetheless, real differences in structure and function have been identified when comparing different kinds of ecological food webs, such as terrestrial vs. aquatic food webs.

Applications of Food Webs

Food webs are constructed to describe species interactions (direct relationships).

The fundamental purpose of food webs is to describe feeding relationship among species in a community. Food webs can be constructed to describe the species interactions. All species in the food webs can be distinguished into basal species (autotrophs, such as plants), intermediate species (herbivores and intermediate level carnivores, such as grasshopper and scorpion) or top predators (high level carnivores such as fox).

These feeding groups are referred as trophic levels. Basal species occupy the lowest trophic level as primary producer. They convert inorganic chemical and use solar energy to generate chemical energy. The second trophic level consists of herbivores. These are first consumers. The remaining trophic levels include carnivores that consume animals at trophic levels below them. The second consumers (trophic level 3) in the desert food web include birds and scorpions, and tertiary consumers making up the fourth trophic level include bird predators and foxes. Grouping all species into different functional groups or tropic levels helps us simplify and understand the relationships among these species.

Food webs can be used to illustrate indirect interactions among species.

Indirect interaction occurs when two species do not interact with each other directly, but influenced by a third species. Species can influence one another in many different ways. One example is the keystone predation are demonstrated by Robert Paine in an experiment conducted in the rocky intertidal zone (Cain et al. 2008; Smith & Smith 2009; Molles 2010). This study showed that predation can influence the competition among species in a food web. The intertidal zone is home to a variety of mussels, barnacles, limpets, and chitons (Paine 1969). All these invertebrate herbivores are preyed upon by the predator starfish Pisaster. Starfish was relatively uncommon in the intertidal zone, and considered less important in the community. When Paine manually removed the starfish from experimental plots while leaving other areas undisturbed as control plots, he found that the number of prey species in the experimental plots dropped from 15 at the beginning of the experiment to 8 (a loss of 7 species) two years after the starfish removal while the total of prey species remained the same in the control plots. He reasoned that in the absence of the predator starfish, several of the mussel and barnacle species (that were superior competitors) excluded the other species and reduced overall diversity in the community (Smith & Smith 2009). Predation by starfish reduced the abundance of mussel and opened up space for other species to colonize and persist. This type of indirect interaction is called keystone predation.

(a) The rocky intertidal zone of the Pacific Northwest coast is inhabited by a variety of species including starfish, barnacles, limpets, chitons, and mussels. (b) A food web of this community shows that the starfish preys on a variety of invertebrate species.

Another interesting study demonstrated indirect interactions among species in both aquatic and terrestrial ecosystems. In a study conducted near Gainesville, Florida, Knight and her colleagues (2009) investigated the effects of fish in ponds on plant seeds production. They measured and compared abundances of both larval and adult dragonfly in and around four ponds that had been stocked with fish and four ponds that lacked fish (Knight et al. 2009). They found that ponds with fish produce fewer larval and adult dragonflies than ponds without fish, as fish prey on larval dragonflies. As dragonfly population decreases, the populations of their prey, including bees, flies, and butterflies, decrease. These prey species are pollinators of the plants. Therefore, flowers in the vicinity of ponds without fish receive fewer pollinator visits than flowers close to ponds stocked with fish. Since the production of seeds is pollen-limited, fewer pollinator visits result in lower seeds production. This study demonstrates, via a complex trophic cascade, that adding fish to a pond improves the reproductive success of a plant on land (Ricklefs 2008).

An interaction food web shows that fish have indirect effects on the populations of several species in and around ponds.

The solid arrows represent direct effects, and the dashed arrows indirect effects; the nature of the effect is indicated by + or -. Fish have indirect effects, through a trophic cascade, on several terrestrial species: dragonfly adults (-), pollinators (+), and plants (+)

Food webs can be used to study bottom-up or top-down control of community structure.

Food webs illustrate energy flow from primary producers to primary consumers (herbivores), and from primary consumers to secondary consumers (carnivores). The structure of food webs suggests that productivity and abundance of populations at any given trophic level are controlled by the productivity and abundance of populations in the trophic level below them (Smith & Smith 2009). This phenomenon is call bottom-up control. Correlations in abundance or productivity between consumers and their resources are considered as evidence for bottom-up control. For example, plant population densities control the abundance of herbivore populations which in turn control the densities of the carnivore populations. Thus, the biomass of herbivores usually increases with primary productivity in terrestrial ecosystems.

Top-down control occurs when the population density of a consumer can control that of its resource, for example, predator populations can control the abundance of prey species (Power 1992). Under top-down control, the abundance or biomass of lower trophic levels depends on effects from consumers at higher trophic levels. A trophic cascade is a type of top-down interaction that describes the indirect effects of predators. In a trophic cascade, predators induce effects that cascade down the food chain and affect biomass of organisms at least two links away (Ricklefs 2008). Nelson Hairston, Frederick Smith and Larry Slobodkin first introduced the concept of top-down control with the frequently quoted "the world is green" proposition (Power 1992; Smith & Smith 2009). They proposed that the world is green because carnivores depress herbivores and keep herbivore populations in check. Otherwise, herbivores would consume most of the vegetation. Indeed, a bird exclusion study demonstrated that there were significantly more insects and leaf damage in plots without birds compared to the control (Marquis & Whelan 1994).

Food webs can be used to reveal different patterns of energy transfer in terrestrial and aquatic ecosystems.

Patterns of energy flow through different ecosystems may differ markedly in terrestrial and aquatic ecosystems (Shurin et al. 2006). Food webs (i.e., energy flow webs) can be used to reveal these differences. In a review paper, Shurin et al. (2006) provided evidence for systematic difference in energy flow and biomass partitioning between producers and herbivores, detritus and decomposers, and higher trophic levels in food webs. A dataset synthesized by Cebrian and colleagues on the fate of carbon fixed by primary productivity across different ecosystems was used to show different patterns in food chains between terrestrial and aquatic ecosystems. On average, the turnover rate of phytoplankton is 10 to 1000 times faster than that of grasslands and forests, thus, less carbon is stored in the living autotroph biomass pool, and producer biomass is consumed by aquatic herbivores at 4 times the terrestrial rate (Cebrian 1999, 2004; Shurin et al. 2006). Herbivores in terrestrial ecosystems are less abundant but decomposers are much more abundant than in phytoplankton dominated aquatic ecosystems. In most terrestrial ecosystems with high standing biomass and relatively low harvest of primary production by herbivores, the detrital food chain is dominant (Smith & Smith 2009). In deep-water aquatic ecosystems, with their low standing biomass, rapid turnover of organisms, and high rate of harvest, the grazing food chain may be dominant.

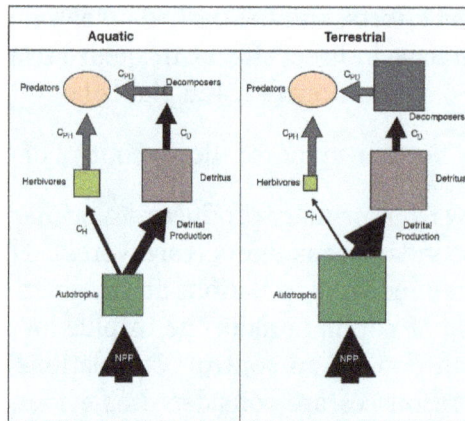

Differences in pathways of carbon flow and pools between aquatic and terrestrial ecosystems.

The thickness of the arrows (flows) and the area of the boxes (pools) correspond to the magnitude. The size of the pools are scaled as log units since the differences cover four orders of magnitude. The C's indicate consumption terms (i.e. CH is consumption by herbivores). Ovals and arrows in grey indicate unknown quantities.

As a diagram tool, food web has been approved to be effective in illustrating species interactions and testing research hypotheses. It will continue to be very helpful for us to understand the associations of species richness/diversity with food web complexity, ecosystem productivity, and stability.

Food Chain

Sample Food Chains

Trophic Level	Grassland Biome	Pond Biome	Ocean Biome
Primary Producer	grass	algae	phytoplankton
Primary Consumer	grasshopper	mosquito larva	zooplankton
Secondary Consumer	rat	dragonfly larva	fish
Tertiary Consumer	snake	fish	seal
Quaternary Consumer	hawk	raccoon	white shark

Every organism needs to obtain energy in order to live. For example, plants get energy from the sun, some animals eat plants, and some animals eat other animals.

A food chain is the sequence of who eats whom in a biological community (an ecosystem) to obtain nutrition. A food chain starts with the primary energy source, usually the sun or boiling-hot deep sea vents. The next link in the chain is an organism that make its own food from the primary energy source -- an example is photosynthetic plants that make their own food from sunlight (using a process called photosynthesis) and chemosynthetic bacteria that make their food energy from chemicals in hydrothermal vents. These are called autotrophs or primary producers.

Next come organisms that eat the autotrophs; these organisms are called herbivores or primary consumers -- an example is a rabbit that eats grass.

The next link in the chain is animals that eat herbivores - these are called secondary consumers -- an example is a snake that eat rabbits.

In turn, these animals are eaten by larger predators -- an example is an owl that eats snakes.

The tertiary consumers are are eaten by quaternary consumers -- an example is a hawk that eats owls. Each food chain end with a top predator, and animal with no natural enemies (like an alligator, hawk, or polar bear).

The arrows in a food chain show the flow of energy, from the sun or hydrothermal vent to a top predator. As the energy flows from organism to organism, energy is lost at each step. A network of many food chains is called a food web.

Ecological Pyramid

An ecological pyramid is a graphical representation of the relationship between different organisms in an ecosystem. Each of the bars that make up the pyramid represents a different trophic level, and their order, which is based on who eats whom, represents the flow of energy. Energy moves up the pyramid, starting with the primary producers, or autotrophs, such as plants and algae at the very bottom, followed by the primary consumers, which feed on these plants, then secondary consumers, which feed on the primary consumers, and so on. The height of the bars should all be the same, but the width of each bar is based on the quantity of the aspect being measured.

Types of Ecological Pyramids

Pyramid of Numbers

This shows the number of organisms in each trophic level without any consideration for their size. This type of pyramid can be convenient, as counting is often a simple task and can be done over the years to observe the changes in a particular ecosystem. However, some types of organisms are difficult to count, especially when it comes to some juvenile forms. Unit: number of organisms.

Pyramid of Biomass

This indicates the total mass of organisms at each trophic level. Usually, this type of pyramid is largest at the bottom and gets smaller going up, but exceptions do exist. The biomass of one trophic level is calculated by multiplying the number of individuals in the trophic level by the average mass of one individual in a particular area. This type of ecological pyramid solves some problems of the pyramid of numbers, as it shows a more accurate representation of the amount of energy contained in each trophic level, but it has its own limitations. For example, the time of year when the data are gathered is very important, since different species have different breeding seasons. Also, since it's usually impossible to measure the mass of every single organism, only a sample is taken, possibly leading to inaccuracies. Unit: g m^{-2} or Kg m^{-2}.

Pyramid of Productivity

The pyramid of productivity looks at the total amount of energy present at each trophic level, as well as the loss of energy between trophic levels. Since this type of representation takes into account the fact that the majority of the energy present at one trophic level will not be available for the next one, it is more accurate than the other two pyramids. This idea is based on Lindeman's Ten Percent Law, which states that only about 10% of the energy in a trophic level will go towards creating biomass. In other words, only about 10% of the energy will go into making tissue, such as stems, leaves, muscles, etc. in the next trophic level. The rest is used in respiration, hunting, and other activities, or is lost to the surroundings as heat. What's interesting, however, is that toxins are passed up the pyramid very efficiently, which means that as we go up the ecological pyramid, the amount of harmful chemicals is more and more concentrated in the organisms' bodies. This is what we call biomagnification.

The pyramid of productivity is the most widely used type of ecological pyramid, and, unlike the two other types, can never be largest at the apex and smallest at the bottom. It's an important type of ecological pyramid because it examines the flow of energy in an ecosystem over time. Unit: J m^{-2} yr^{-1}, where Joule is the unit for energy, which can be interchanged by other units of energy such as Kilojoule, Kilocalorie, and calorie.

While a productivity pyramid always takes an upright pyramid shape, number pyramids are sometimes inverted, or don't take the shape of an actual pyramid at all. To demonstrate, let's take an oak tree, which can feed millions of oakworms. If we consider this ecosystem as our focus, then the producers' level (one tree) will end up much smaller than the primary consumers' level (millions of insects). This is less likely to occur in biomass pyramids, but is not impossible. The pyramids below show the different types of pyramids and the shapes they can have in different ecosystems.

Ecological Pyramid Examples

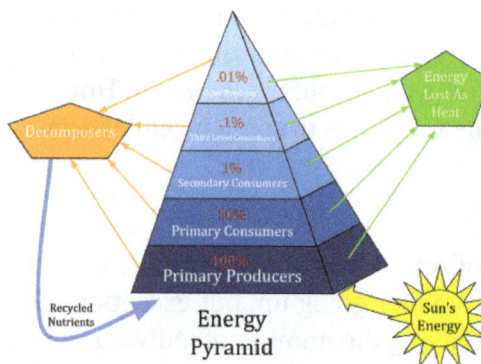

The diagram below is an example of a productivity pyramid, otherwise called an energy pyramid.

The sun has been included in this diagram, as it's the main source of all energy, as well the decomposers, like bacteria and fungi, which can acquire nutrients and energy from all trophic levels by breaking down dead or decaying organisms. As shown, the nutrients then go back into the soil and are taken up by plants.

The loss of energy to the surroundings is also shown in this diagram, and the total energy transfer has been calculated. We start off with the total amount of energy that the primary producers contain, which is indicated by 100%. As we go up one level, 90% of that energy is used in ways other than to create flesh. What the primary consumers end up with is just 10% of the starting energy, and, 10% of that 10% is lost in the transfer to the next level. That's 1%, and so on. The predators at the apex, then, will only receive 0.01% of the starting energy! This inefficiency in the system is the reason why productivity pyramids are always upright.

Function of Ecological Pyramid

An ecological pyramid not only shows us the feeding patterns of organisms in different ecosystems, but can also give us an insight into how inefficient energy transfer is, and show the influence that a change in numbers at one trophic level can have on the trophic levels above and below it. Also, when data are collected over the years, the effects of the changes that take place in the environment on the organisms can be studied by comparing the data. If an ecosystem's conditions are found to be worsening over the years because of pollution or overhunting by humans, action can be taken to prevent further damage and possibly reverse some of the present damage.

Ecological Efficiency

Ecological efficiency measures the success of production activities in minimizing a required natural flow.

A change in ecological efficiencies can shift budget shares and share limits, and thus ecological limits and target quantities.

The concept is also required to establish target rates for natural flows. Note that, although the other limits discussed are based on biological flows, ecological efficiency applies to all natural flows, and can therefore be specified for nonrenewables as well.

Ecological efficiency (E_E) is the relationship between a natural flow and its associated output, and is expressed as a ratio — the output's quantity divided by the amount of the flow used in its production:

$$E_E = Q/flow$$

If an output incorporates multiple natural flows, it will have multiple ecological efficiencies associated with it.

Because ecological efficiency is a technical measure (its definition excludes health), it is relevant to production in general, and can therefore be applied to any stage of the output life cycle.

For instance, even though a printing press is an intermediate output and thus lacks potential value, we can calculate the ecological efficiencies for the natural flows used in its production.

Ecological efficiency is a ratio of mixed dimensions. The numerator is a unit quantity of an output, and the denominator reflects the material nature of the associated flow. Following are three examples of such ratios:

- One house/Board feet of lumber

- One construction beam/Pounds of iron

- 1,000 hours of consulting services/Kilograms of greenhouse gases

The mixed nature of this ratio means that ecological efficiencies are commensurable only if the outputs are the same and the natural flows are of the same types. For example, it is possible to compare the ecological efficiencies of two house-building methods by citing how many board feet of lumber each requires for a standard-size house.

However, it is not possible to compare these ecological efficie cies with those for the construction of a commercial building that uses steel.

The effects of changes in ecological efficiencies on share limits (for a single final output) and budget limits (for the economy as a whole) are depicted in the following figure.

Ecological efficiency and share/budget limits

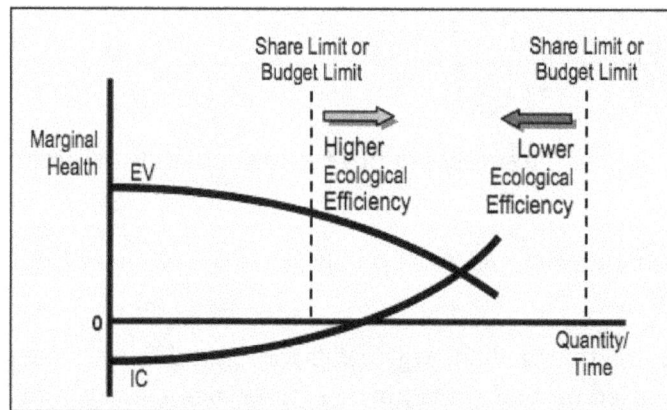

An increase in ecological efficiency will increase a share limit or budget limit, thus permitting higher output quantities. A decrease has the opposite effect. "EV" and "IC" indicate effectual value and input cost respectively.

Consider a single output first. If ecological efficiency rises for a biological flow, less of the flow is required per unit of output, which means that more of the output can be produced within its budget share.

The output's associated share limit thus increases — that is, it shifts to the right. If this share limit is also the ecological limit, the ecological limit increases. If the ecological limit is also the target quantity, the target quantity increases as well.

The reverse sequence applies if ecological efficiency falls. For an economy's total outputs, a rise in ecological efficiency means that the associated budget limit shifts to the right. If ecological efficiency falls, the budget limit shifts to the left. The impacts on the ecological limit and target scale are parallel to those for a single output.

It is important to recognize that increased ecological efficiency is the only way to increase share limits and budget limits. These limits are based entirely on the physical world, and therefore cannot be affected by economic factors such value, cost, or population levels.

On a graph the limits could therefore be imagined as fixed, with the other quantities shifting around them, unless technical changes occur that modify their associated ecological efficiencies.

Ecological efficiencies can be illustrated by the example of houses. If we stop clearcut logging and choose a more environmentally benign method of harvesting wood, habitat destruction per unit of wood will decrease.

This will constitute an increase in ecological efficiency for this flow, and will permit us to extract more wood and thus to build more houses.

If we are sufficiently successful in reducing habitat destruction, the ecological limit will shift to the wood budget. If this occurs, our efficiency efforts should be redirected — we should now focus on using less wood per house, building smaller houses, emphasizing multiple-occupancy residences over single-family houses, or perhaps moving away from wood to a construction material with less environmental impact.

Efforts such as this will continually shift the ecological limit to the next higher share limit, until the ecological limit is higher than the optimum quantity and we can rationally produce up to the output's economic limit.

A few additional points should be made regarding ecological efficiencies:

First, if an output is recycled rather than discarded, some of the natural resources it contains will be recovered. This will effectively reduce the natural flow requirement per unit of output quantity. Recycling is thus one way to increase ecological efficiency.

Second, ecological efficiency applies not only to scarce resources (those obtained through economic production, such as metals and oil), but also to resources that are widely available without production, such as air, sunlight, and water.

For example, a solar panel than uses less sunlight per unit of electricity is more ecologically efficient than one that uses more, even though sunlight is not the result of production. The same can be said for a windmill with respect to wind requirements.

The unconditional maximization of ecological efficiencies is necessary if we are to delay the onset of scarcity, thus postponing the day when painful choices must be made with respect to resource allocation.

Third, an increase in ecological efficiencies will typically lower natural cost. If less of a pollutant is generated per unit of output, the human damage from this production will likely be reduced as well.

Of course this reduced cost can easily be negated by an increase in output quantity. This is called the rebound effect or Jevons paradox, and arises when people consume more of the same output with the money they save from higher efficiencies. Although this is always a potential danger, it should not be a major issue for an ENL-driven economy.

Last, ecological efficiencies complicate the ranking of production facilities, which in ENL is based solely on potential gains. However, it is possible that facility A achieves higher potential gains than facility B, but that B has higher ecological efficiencies than A.

In such cases, which production facility should be preferred?

Unfortunately, no analytical method appears to exist here. Potential gains relate largely to present health, whereas ecological efficiency relates largely to the natural conditions for future health. Such temporally separated quantities are difficult to compare, and social judgment must therefore determine which production facility should be preferred.

ENL's definition of ecological efficiency is strongly reminiscent of the well-known "IPAT" equation. This was developed in the 1970s during debates about the factors involved in ecological impact.

Written out in full, the IPAT formula is this:

Impact = Population * Affluence * Technology

This means that the level of ecological impact is determined by population, affluence (per capita consumption), and the level of technology. If we interpret the latter as efficiency, it must appear as the denominator. The formula then becomes:

$I = PA/T$

Thus, ecological impact rises if population increases or per capita consumption increases; it drops if technical efficiency increases.

Compare this to ENL's formula for ecological efficiency: $E_E = Q/\text{flow}$. This can be rearranged as follows:

$\text{Flow} = Q/E_E$

The terms in these two formulas are equivalent: ecological impact is the result of natural flows, population multiplied by affluence results in output quantity, and technical efficiency is equivalent to ecological efficiency.

The formulas are thus consistent with each other, reflecting the fact that they are different approaches to the same underlying issues.

Primary Production

"Primary production" refers to energy fixed by plants. The total amount of energy fixed is usually called "gross production." A certain fraction of gross production is used in respiration by the plants; the remainder appears as new biomass or "net primary production." Thus for a single plant or a community of green plants:

Net Primary Production = Gross Production – Respiration (of Autotrophs)

Similar relationships occur in ecosystems except that the organic matter and respiration of heterotrophs must be included. The increase in total organic matter is "net ecosystem production"; respiration is the total respiration of the green plants (autotrophs) and the animal community and decay organisms (heterotrophs). Gross production is of course identical to that of the plant community. Thus for an ecosystem:

Net Ecosystem Production = Gross Production – Respiration (of Autotrophs and Heterotrophs).

References

- Leith, H.; Whittaker, R.H. (1975). Primary Productivity of the Biosphere. New York: Springer-Verlag. ISBN 0-387-07083-4

- Reichstein, Markus; Falge, Eva; Baldocchi, Dennis; Papale, Dario; Aubinet, Marc; Berbigier, Paul; et al. (2005). "On the separation of net ecosystem exchange into assimilation and ecosystem respiration: review and improved algorithm". Global Change Biology. 11 (9): 1424–1439. Bibcode:2005GCBio..11.1424R. doi:10.1111/j.1365-2486.2005.001002.x. ISSN 1354-1013

- Paine, R. T. (1980). "Food webs: Linkage, interaction strength and community infrastructure". Journal of Animal Ecology. 49 (3): 666–685. doi:10.2307/4220. JSTOR 4220

- "Net Primary Productivity : Global Maps". earthobservatory.nasa.gov. 26 March 2018. Retrieved 26 March2018

- Tscharntke, T.; Hawkins, B., A., eds. (2002). Multitrophic Level Interactions. Cambridge: Cambridge University Press. p. 282. ISBN 0-521-79110-3

- Martin, J. H.; Fitzwater, S. E. (1988). "Iron-deficiency limits phytoplankton growth in the Northeast Pacific Subarctic". Nature. 331 (6154): 341–343. Bibcode:1988Natur.331..341M. doi:10.1038/331341a0

- Dunne, J. A.; Williams, R. J.; Martinez, N. D.; Wood, R. A.; Erwin, D. H.; Dobson, Andrew P. (2008). "Compilation and Network Analyses of Cambrian Food Webs". PLOS Biology. 6 (4): e102. doi:10.1371/journal.pbio.0060102

- Sigman, D.M.; Hain, M.P. (2012). "The Biological Productivity of the Ocean" (PDF). Nature Education Knowledge. 3 (6): 1–16. Retrieved 2015-06-01

- Odum, E. P.; Barrett, G. W. (2005). Fundamentals of Ecology (5th ed.). Brooks/Cole, a part of Cengage Learning. ISBN 0-534-42066-4. Archived from the original on 2011-08-20

- Clark, D.A.; Brown, S.; Kicklighter, D.W.; Chambers, J.Q.; Thomlinson, J.R.; Ni, J. (2001). "Measuring net primary production in forests: Concepts and field methods" (Scholar search). Ecological Applications. 11 (2): 356–370. doi:10.1890/1051-0761(2001)011[0356:MNPPIF]2.0.CO;2. ISSN 1051-0761

- Montoya, J. M.; Solé, R. V. (2002). "Small world patterns in food webs" (PDF). Journal of Theoretical Biology. 214 (3): 405–412. arXiv:cond-mat/0011195. doi:10.1006/jtbi.2001.2460. Archived from the original (PDF) on 2011-09-05

- Rickleffs, Robert, E. (1996). The Economy of Nature. University of Chicago Press. p. 678. ISBN 0-7167-3847-3

- Jump up^ Luz and Barkan, B; Barkan, E (2000). "Assessment of oceanic productivity with the triple-isotope composition of dissolved oxygen". Science. 288 (5473): 2028–2031. Bibcode:2000Sci...288.2028L. doi:10.1126/science.288.5473.2028. PMID 10856212

Chapter 3

Population Ecology

Population ecology is a sub-field of ecology. It studies the interactions between species populations and the environment. The concepts of population growth, population dynamics, abundance and r/K selection theory are fundamental to the development of population ecology. All such topics have been extensively discussed in this chapter.

Population ecology, study of the processes that affect the distribution and abundance of animal and plant populations.

A population is a subset of individuals of one species that occupies a particular geographic area and, in sexually reproducing species, interbreeds. The geographic boundaries of a population are easy to establish for some species but more difficult for others. For example, plants or animals occupying islands have a geographic range defined by the perimeter of the island. In contrast, some species are dispersed across vast expanses, and the boundaries of local populations are more difficult to determine. A continuum exists from closed populations that are geographically isolated from, and lack exchange with, other populations of the same species to open populations that show varying degrees of connectedness.

Population Factors

Ecologists describe the organisms of populations in several different ways. The distribution of a population is the total area that population covers. The abundance of a population is the number of individuals within that population. Ecologists may also define the number of individuals within a certain space, which is the density of the population.

Ecologists also identify the age structure or sex ratio of a population. The age structure describes the number of individuals in different age classes, while the sex ratio describes the proportion of males to females in that population.

Population Growth

Within any population, individuals are born and individuals die. If there are more individuals being born than dying, the population grows in size, while if more individuals are dying than being born, the population shrinks. Individuals may also enter or leave a population, which is referred to as immigration and emigration.

To better understand population growth, ecologists have created models to study how birth, death, immigration, and emigration affect population size. The simplest model is called the exponential growth model. It says that the change in population size is exponential, or growing at an increasing rate. This is not a very realistic model because most populations do not continue to grow without slowing down.

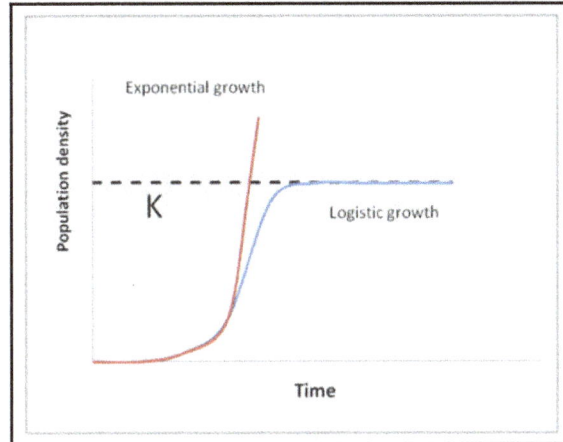

Fundamentals

Terms used to describe natural groups of individuals in ecological studies	
Term	Definition
Species population	All individuals of a species.
Metapopulation	A set of spatially disjunct populations, among which there is some immigration.
Population	A group of conspecific individuals that is demographically, genetically, or spatially disjunct from other groups of individuals.
Aggregation	A spatially clustered group of individuals.
Deme	A group of individuals more genetically similar to each other than to other individuals, usually with some degree of spatial isolation as well.
Local population	A group of individuals within an investigator-delimited area smaller than the geographic range of the species and often within a population (as defined above). A local population could be a disjunct population as well.
Subpopulation	An arbitrary spatially delimited subset of individuals from within a population (as defined above).

The most fundamental law of population ecology is Thomas Malthus' exponential law of population growth.

A population will grow (or decline) exponentially as long as the environment experienced by all individuals in the population remains constant.

Thomas Robert Malthus

This principle in population ecology provides the basis for formulating predictive theories and tests that follow:

Simplified population models usually start with four key variables (four demographic processes) including death, birth, immigration, and emigration. Mathematical models used to calculate changes in population demographics and evolution hold the assumption (or null hypothesis) of no external influence. Models can be more mathematically complex where "…several competing hypotheses are simultaneously confronted with the data." For example, in a closed system where immigration and emigration does not take place, the rate of change in the number of individuals in a population can be described as:

$$\frac{dN}{dT} = B - D = bN - dN = (b-d)N = rN,$$

where N is the total number of individuals in the population, B is the raw number of births, D is the raw number of deaths, b and d are the per capita rates of birth and death respectively, and r is the per capita average number of surviving offspring each individual has. This formula can be read as the rate of change in the population (dN/dT) is equal to births minus deaths (B - D).

Using these techniques, Malthus' population principle of growth was later transformed into a mathematical model known as the logistic equation:

$$\frac{dN}{dT} = aN\left(1 - \frac{N}{K}\right),$$

where N is the biomass density, a is the maximum per-capita rate of change, and K is the carrying capacity of the population. The formula can be read as follows: the rate of change in the population (dN/dT) is equal to growth (aN) that is limited by carrying capacity (1-N/K). From these basic mathematical principles the discipline of population ecology expands into a field of investigation that queries the demographics of real populations and tests these results against the statistical models. The field of population ecology often uses data on life history and matrix algebra to develop projection matrices on fecundity and survivorship. This information is used for managing wildlife stocks and setting harvest quotas.

Geometric Populations

Operophtera brumata (Winter moth) populations are geometric.

The population model below can be manipulated to mathematically infer certain properties of geometric populations. A population with a size that increases geometrically is a population where generations of reproduction do not overlap. In each generation there is an effective population size denoted as N_e which constitutes the number of individuals in the population that are able to reproduce and will reproduce in any reproductive generation in concern. In the population model below it is assumed that N is the effective population size.

Assumption 01: $N_e = N$

$N_{t+1} = N_t + B_t + I_t - D_t - E_t$

Term	Definition
N_{t+1}	Population size in the generation after generation $_t$. This may be the current generation or the next (upcoming) generation depending on the situation in which the population model is used.
N_t	Population size in generation $_t$.
B_t	Sum (Σ) of births in the population between generations $_t$ and $_{t+1}$. Also known as raw birth rate.
I_t	Sum (Σ) of immigrants moving into the population between generations $_t$ and $_{t+1}$. Also known as raw immigration rate.
D_t	Sum (Σ) of deaths in the population between generations $_t$ and $_{t+1}$. Also known as raw death rate.
E_t	Sum (Σ) of emigrants moving out of the population between generations $_t$ and $_{t+1}$. Also known as raw emigration rate.

The general difference between populations that grow exponentially and geometrically.

Geometric populations grow in reproductive generations between intervals of abstinence from reproduction. Exponential populations grow without designated periods for reproduction. Reproduction is a continuous process and generations of reproduction overlap. This graph illustrates two hypothetical populations - one population growing periodically (and therefore geometrically) and the other population growing continuously (and therefore exponentially). The populations in the graph have a doubling time of 1 year. The populations in the graph are hypothetical. In reality, the doubling times differ between populations.

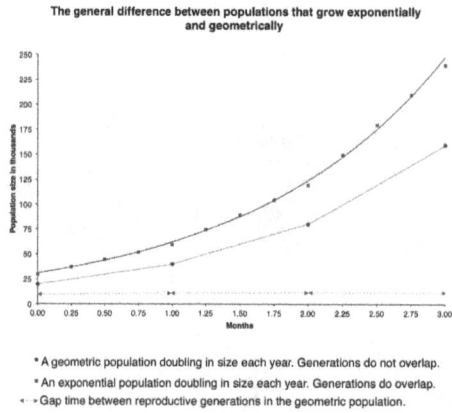

The general difference between populations that grow exponentially and geometrically

* A geometric population doubling in size each year. Generations do not overlap.
* An exponential population doubling in size each year. Generations do overlap.
⬩ ⬩ Gap time between reproductive generations in the geometric population.

Assumption 02: There is no migration to or from the population (N)

$$I_t = E_t = 0$$

$$N_{t+1} = N_t + B_t - D_t$$

The raw birth and death rates are related to the per capita birth and death rates:

$$B_t = b_t \times N_t$$

$$D_t = d_t \times N_t$$

$$b_t = B_t / N_t$$

$$d_t = D_t / N_t$$

Term	Definition
b_t	Per capita birth rate.
d_t	Per capita death rate.

Therefore:

$$N_{t+1} = N_t + (b_t \times N_t) - (d_t \times N_t)$$

Assumption 03: b_t and d_t are constant (i.e. they don't change each generation).

$$N_{t+1} = N_t + (bN_t) - (dN_t)$$

Term	Definition
b	Constant per capita birth rate.
d	Constant per capita death rate.

Take the term N_t out of the brackets.

$$N_{t+1} = N_t + (b - d)N_t$$

$$b - d = R$$

Term	Definition
R	Geometric rate of increase.

$$N_{t+1} = N_t + RN_t$$

$$N_{t+1} = (N_t + RN_t)$$

Take the term N_t out of the brackets again.

$$N_{t+1} = (1 + R)N_t$$

$$1 + R = \lambda$$

Term	Definition
λ	Finite rate of increase.

$$N_{t+1} = \lambda N_t$$

At $_{t+1}$	$N_{t+1} = \lambda N_t$
At $_{t+2}$	$N_{t+2} = \lambda N_{t+1} = \lambda\lambda N_t = \lambda^2 N_t$
At $_{t+3}$	$N_{t+3} = \lambda N_{t+2} = \lambda\lambda N_{t+1} = \lambda\lambda\lambda N_t = \lambda^3 N_t$
At $_{t+4}$	$N_{t+4} = \lambda N_{t+3} = \lambda\lambda N_{t+2} = \lambda\lambda\lambda N_{t+1} = \lambda\lambda\lambda\lambda N_t = \lambda^4 N_t$
At $_{t+5}$	$N_{t+5} = \lambda N_{t+4} = \lambda\lambda N_{t+3} = \lambda\lambda\lambda N_{t+2} = \lambda\lambda\lambda\lambda N_{t+1} = \lambda\lambda\lambda\lambda\lambda N_t = \lambda^5 N_t$

Therefore:

$$N_{t+1} = \lambda^t N_t$$

Term	Definition
λ^t	Finite rate of increase raised to the power of the number of generations (e.g. for $_{t+2}$ [two generations] $\rightarrow \lambda^2$, for $_{t+1}$ [one generation] $\rightarrow \lambda^1 = \lambda$, and for $_t$ [before any generations - at time zero] $\rightarrow \lambda^0 = 1$

Doubling Time of Geometric Populations

G. stearothermophilus has a shorter doubling time (td) than E. coli and N. meningitidis. Growth rates of 2 bacterial species will differ by unexpected orders of magnitude if the doubling times of the 2 species differ by even as little as 10 minutes. In eukaryotes such as animals, fungi, plants, and protists, doubling times are much longer than in bacteria. This reduces the growth rates of eukaryotes in comparison to Bacteria. G. stearothermophilus, E. coli, and N. meningitidis have 20 minute, 30 minute, and 40 minute doubling times under optimal conditions respectively. If bacterial populations could grow indefinitely (which they do not) then the number of bacteria in each species would approach infinity (∞). However, the percentage of G. stearothermophilus bacteria out of all the bacteria would approach 100% whilst the percentage of E. coli and N. meningitidis combined out of all the bacteria would approach 0%. This graph is a simulation of this hypothetical scenario. In reality, bacterial populations do not grow indefinitely in size and the 3 species require different optimal conditions to bring their doubling times to minima.

Time in minutes	% that is G. stearothermophilus
30	44.4%
60	53.3%
90	64.9%
120	72.7%
$\rightarrow\infty$	100%

Time in minutes	% that is E. coli
30	29.6%
60	26.7%
90	21.6%
120	18.2%
→∞	0.00%

Time in minutes	% that is N. meningitidis
30	25.9%
60	20.0%
90	13.5%
120	9.10%
→∞	0.00%

Bacterial populations are exponential (or, more correctly, logistic) instead of geometric.
Nevertheless, doubling times are applicable to both types of populations.

The doubling time of a population is the time required for the population to grow to twice its size. We can calculate the doubling time of a geometric population using the equation: $N_{t+1} = \lambda^t N_t$ by exploiting our knowledge of the fact that the population (N) is twice its size (2N) after the doubling time.

$$2N_{td} = \lambda^t_d \times N_t$$

Term	Definition
t_d	Doubling time.

$$\lambda^t_d = 2N_{td} / N_t$$

$$\lambda^t_d = 2$$

The doubling time can be found by taking logarithms. For instance:

$$t_d \times \log_2(\lambda) = \log_2(2)$$

$$\log_2(2) = 1$$

$$t_d \times \log_2(\lambda) = 1$$

$$t_d = 1 / \log_2(\lambda)$$

Or:

$$t_d \times \ln(\lambda) = \ln(2)$$

$$t_d = \ln(2) / \ln(\lambda)$$

$$t_d = 0.693... / \ln(\lambda)$$

Therefore:

$$t_d = 1 / \log_2(\lambda) = 0.693... / \ln(\lambda)$$

Half-life of Geometric Populations

The half-life of a population is the time taken for the population to decline to half its size. We can calculate the half-life of a geometric population using the equation: $N_{t+1} = \lambda^t N_t$ by exploiting our knowledge of the fact that the population (N) is half its size (0.5N) after a half-life.

$$0.5N_{t1/2} = \lambda^t_{1/2} \times N_t$$

Term	Definition
$t_{1/2}$	Half-life.

$$\lambda^t_{1/2} = 0.5N_{t1/2} / N_t$$

$$\lambda^t_{1/2} = 0.5$$

The half-life can be calculated by taking logarithms.

$$t_{1/2} = 1 / \log_{0.5}(\lambda) = \ln(0.5) / \ln(\lambda)$$

Geometric (R) and Finite (λ) Growth Constants

Geometric (R) Growth Constant

$$R = b - d$$

$$N_{t+1} = N_t + RN_t$$

$$N_{t+1} - N_t = RN_t$$

$$N_{t+1} - N_t = \Delta N$$

Term	Definition
ΔN	Change in population size between two generations (between generation $_{t+1}$ and $_t$).

$$\Delta N = RN_t$$

$$\Delta N/N_t = R$$

Finite (λ) growth constant

$$1 + R = \lambda$$

$$N_{t+1} = \lambda N_t$$

$$\lambda = N_{t+1} / N_t$$

Mathematical Relationship Between Geometric and Exponential Populations

In geometric populations, R and λ represent growth constants. In exponential populations however, the intrinsic growth rate, also known as intrinsic rate of increase (r) is the relevant growth constant. Since generations of reproduction in a geometric population do not overlap (e.g. reproduce once a year) but do in an exponential population, geometric and exponential populations are usually considered to be mutually exclusive. However, geometric constants and exponential constants share the mathematical relationship below.

The growth equation for exponential populations is

$$N_t = N_o e^{rt}$$

Term	Definition
e	Euler's number - A universal constant often applicable in exponential equations.
r	intrinsic growth rate - also known as intrinsic rate of increase.

Leonhard Euler was the mathematician who established the universal constant 2.71828... also known as Euler's number or e.

Assumption: N_t (of a geometric population) = N_t (of an exponential population).

Therefore:

$$N_o e^{rt} = N_o \lambda^t$$

N_o cancels on both sides.

$$N_o e^{rt} / N_o = \lambda^t$$

$$e^{rt} = \lambda^t$$

Take the natural logarithms of the equation. Using natural logarithms instead of base 10 or base 2 logarithms simplifies the final equation as $\ln(e) = 1$.

$$rt \times \ln(e) = t \times \ln(\lambda)$$

Term	Definition
ln	Natural logarithm - in other words $\ln(y) = \log_e(y) = x$ = the power (x) that e needs to be raised to (e^x) to give the answer y.
	In this case, $e^1 = e$ therefore $\ln(e) = 1$.

$$rt \times 1 = t \times \ln(\lambda)$$

$$rt = t \times \ln(\lambda)$$

t cancels on both sides.

$$rt / t = \ln(\lambda)$$

The results:

$$r = \ln(\lambda)$$

and

$$e^r = \lambda$$

R/K Selection

An important concept in population ecology is the r/K selection theory. The first variable is r (the intrinsic rate of natural increase in population size, density independent) and the second variable is K (the carrying capacity of a population, density dependent). An r-selected species (e.g., many kinds of insects, such as aphids) is one that has high rates of fecundity, low levels of parental investment in the young, and high rates of mortality before individuals reach maturity. Evolution favors productivity in r-selected species. In contrast, a K-selected species (such as humans) has low rates of fecundity, high levels of parental investment in the young, and low rates of mortality as individuals mature. Evolution in K-selected species favors efficiency in the conversion of more resources into fewer offspring.

Metapopulation, in ecology, a regional group of connected populations of a species. For a given species, each metapopulation is continually being modified by increases (births and immigrations) and decreases (deaths and emigrations) of individuals, as well as by the emergence and dissolution of local populations contained within it. As local populations of a given species fluctuate in size, they become vulnerable to extinction during periods when their numbers are low. Extinction of local populations is common in some species, and the regional persistence of such species is dependent on the existence of a metapopulation. Hence, elimination of much of the metapopulation structure of some species can increase the chance of regional extinction of species.

The structure of metapopulations varies among species. In some species one population may be particularly stable over time and act as the source of recruits into other, less stable populations. For example, populations of the checkerspot butterfly (Euphydryas editha) in California have a metapopulation structure consisting of a number of small satellite populations that surround a large source population on which they rely for new recruits. The satellite populations are too small and fluctuate too much to maintain themselves indefinitely. Elimination of the source population from this metapopulation would probably result in the eventual extinction of the smaller satellite populations.

In other species, metapopulations may have a shifting source. Any one local population may temporarily be the stable source population that provides recruits to the more unstable surrounding populations. As conditions change, the source population may become unstable, as when disease increases locally or the physical environment deteriorates. Meanwhile, conditions in another population that had previously been unstable might improve, allowing this population to provide recruits.

Population Size

Demography: Describing Populations and How they Change

In many cases, ecologists aren't studying people in towns and cities. Instead, they're studying various kinds of plant, animal, fungal, and even bacterial populations. The statistical study of any population, human or otherwise, is known as demography.

Why is demography important? Populations can change in their numbers and structure—for example age and sex distribution—for various reasons. These changes can affect how the population interacts with its physical environment and with other species.

By tracking populations over time, ecologists can see how these populations have changed and may be able to predict how they're likely to change in the future. Monitoring the size and structure of populations can also help ecologists manage populations—for example, by showing whether conservation efforts are helping an endangered species increase in numbers.

In this article, we'll begin our journey through demographics by looking at the concepts of population size, density, and distribution. We'll also explore some methods ecologists use to determine these values for populations in nature.

Population Size and Density

To study the demographics of a population, we'll want to start off with a few baseline measures. One is simply the number of individuals in the population, or population size—NNNN. Another is the population density, the number of individuals per area or volume of habitat.

Size and density are both important in describing the current status of the population and, potentially, for making predictions about how it could change in the future:

- Larger populations may be more stable than smaller populations because they're likely to have greater genetic variability and thus more potential to adapt to changes in the environment through natural selection.

- A member of a low-density population—where organisms are sparsely spread out—might have more trouble finding a mate to reproduce with than an individual in a high-density population.

Measuring Population Size

To find the size of a population, can't we just count all the organisms in it? Ideally, yes! But in many real-life cases, this isn't possible. For instance, would you want to try and count every single grass plant in your lawn? Or every salmon in, say, Lake Ontario, which is 393 cubic miles in volume?1^11start superscript, 1, end superscript Counting all the organisms in a population may be too expensive in terms of time and money, or it may simply not be possible.

For these reasons, scientists often estimate a population's size by taking one or more samples from the population and using these samples to make inferences about the population as a whole. A variety of methods can be used to sample populations to determine their size and density. Here, we'll look at two of the most important: the quadrat and mark-recapture methods.

Quadrat Method

For immobile organisms such as plants—or for very small and slow-moving organisms—plots called quadrats may be used to determine population size and density. Each quadrat marks off an area of the same size—typically, a square area—within the habitat. A quadrat can be made by staking out an area with sticks and string or by using a wood, plastic, or metal square placed on the ground, as shown in the picture below.

After setting up quadrats, researchers count the number of individuals within the boundaries of each one. Multiple quadrat samples are performed throughout the habitat at several random locations, which ensures that the numbers recorded are representative for the habitat overall. In the end, the data can be used to estimate the population size and population density within the entire habitat.

Mark-recapture Method

For organisms that move around, such as mammals, birds, or fish, a technique called the mark-recapture method is often used to determine population size. This method involves capturing a sample of animals and marking them in some way—for instance, using tags, bands, paint, or other body markings, as shown below. Then, the marked animals are released back into the environment and allowed to mix with the rest of the population.

Later, a new sample is collected. This new sample will include some individuals that are marked—recaptures—and some individuals that are unmarked. Using the ratio of marked to unmarked individuals, scientists can estimate how many individuals are in the total population.

Example: Using the Mark-recapture Method

Let's say we want to find the size of a deer population. Suppose that we capture 80 deer, tag them, and release them back into the forest. After some time has passed—allowing the marked deer to thoroughly mix with the rest of the population—we come back and capture another 100 deer. Out of these deer, we find that 20 are already marked.

If 20 out of 100 deer are marked, this would suggest that marked deer—which we know are 80 in number—make up 20% of the population. Using this information, we can formulate the following relationship:

$$\frac{\text{number marked first catch}\left(M\right)}{\text{total population}\left(N\right)} = \frac{\text{number marked second catch }(x)}{\text{total number of second catch }(n)}$$

$$\frac{M}{N} = \frac{x}{n}$$

Next, we rearrange the equation:

$$N = \frac{nM}{x}$$

And finally, we plug in the values from the deer example:

$$N = \frac{(100 \text{ total second catch})(80 \text{ marked first catch})}{(20 \text{ marked second catch})} = 400 \, \text{deer}$$

This approach isn't always perfect. Some animals from the first catch may learn to avoid capture in the second round, inflating population estimates. Alternatively, the same animals may preferentially be retrapped—especially if a food reward is offered—resulting in an underestimate of population size. Also, some species may be harmed by the marking technique, reducing their survival. The approach also assumes that animals don't die, get born, leave, or enter the population during the period of the study.

Alternative approaches to determine population size include electronic tracking of animals tagged with radio transmitters and use of data from commercial fishing and trapping operations.

Species Distribution

Often, in addition to knowing the number and density of individuals in an area, ecologists will also want to know their distribution. Species dispersion patterns—or distribution patterns—refer to how the individuals in a population are distributed in space at a given time.

The individual organisms that make up a population can be more or less equally spaced, dispersed randomly with no predictable pattern, or clustered in groups. These are known as uniform, random, and clumped dispersion patterns, respectively.

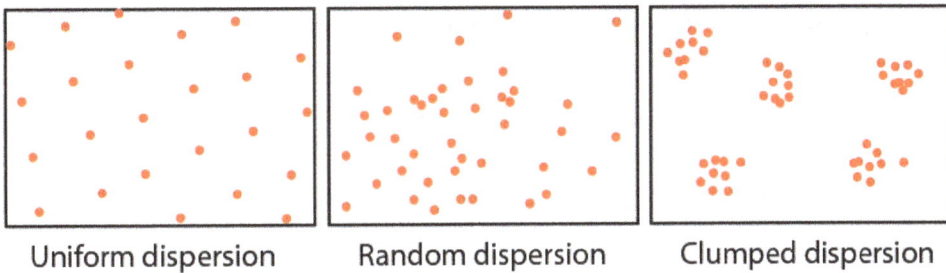

Uniform dispersion   Random dispersion   Clumped dispersion

- *Uniform dispersion.* In uniform dispersion, individuals of a population are spaced more or less evenly. One example of uniform dispersion comes from plants that secrete toxins to inhibit growth of nearby individuals—a phenomenon called allelopathy. We can also find uniform dispersion in animal species where individuals stake out and defend territories.

- *Random dispersion.* In random dispersion, individuals are distributed randomly, without a predictable pattern. An example of random dispersion comes from dandelions and other plants that have wind-dispersed seeds. The seeds spread widely and sprout where they happen to fall, as long as the environment is favorable—has enough soil, water, nutrients, and light.

- *Clumped dispersion.* In a clumped dispersion, individuals are clustered in groups. A clumped dispersion may be seen in plants that drop their seeds straight to the ground—such as oak trees—or animals that live in groups—schools of fish or herds of elephants. Clumped dispersions also happen in habitats that are patchy, with only some patches suitable to live in.

As you can see from these examples, dispersion of individuals in a population provides more information about how they interact with each other—and with their environment—than a simple density measurement.

Effective Population Size

Effective population size is the number of individuals in a population who contribute offspring to the next generation.

In an ecological sense, the size of a population can be measured by simply counting the number of adults in a locality. However, for the theory of population genetics what matters is the chance that two copies of a gene will be sampled as the next generation is produced, and this is affected by the breeding structure of the population.

In a population of size N there will be 2N genes. The correct interpretation of N for the theoretical equations is that N has been correctly measured when the chance of drawing two copies of the same gene is $(1/2N)^2$. If we draw two genes from a population at a locality, we may be more likely for various reasons to get two copies of the same gene than would be implied by the naive ecological measure of population size.

Population geneticists therefore often write N_e (for 'effective' population size) in the equations, rather than N. In practice, effective population sizes are usually lower than ecologically observed population sizes.

Types of Effective Population Size

Depending on the quantity of interest, effective population size can be defined in several ways. Ronald Fisher and Sewall Wright originally defined it as "the number of breeding individuals in an idealised population that would show the same amount of dispersion of allele frequencies under random genetic drift or the same amount of inbreeding as the population under consideration". More generally, an effective population size may be defined as the number of individuals in an idealised population that has a value of any given population genetic quantity that is equal to the value of that quantity in the population of interest. The two population genetic quantities identified by Wright were the one-generation increase in variance across replicate populations (variance effective population size) and the one-generation change in the inbreeding coefficient (inbreeding effective population size). These two are closely linked, and derived from F-statistics, but they are not identical.

Today, the effective population size is usually estimated empirically with respect to the sojourn or coalescence time, estimated as the within-species genetic diversity divided by the mutation rate, yielding a coalescent effective population size. Another important effective population size is the selection effective population size $1/s_{critical}$, where $s_{critical}$ is the critical value of the selection coefficient at which selection becomes more important than genetic drift.

Empirical Measurements

In Drosophila populations of census size 16, the variance effective population size has been measured as equal to 11.5. This measurement was achieved through studying changes in the frequency of a neutral allele from one generation to another in over 100 replicate populations.

For coalescent effective population sizes, a survey of publications on 102 mostly wildlife animal and plant species yielded 192 N_e/N ratios. Seven different estimation methods were used in the surveyed studies. Accordingly, the ratios ranged widely from 10^{-6} for Pacific oysters to 0.994 for humans, with an average of 0.34 across the examined species. A genealogical analysis of human hunter-gatherers (Eskimos) determined the effective-to-census population size ratio for haploid (mitochondrial DNA, Y chromosomal DNA), and diploid (autosomal DNA) loci separately: the ratio of the effective to the census population size was estimated as 0.6–0.7 for autosomal and X-chromosomal DNA, 0.7–0.9 for mitochondrial DNA and 0.5 for Y-chromosomal DNA.

Variance Effective Size

References missing In the Wright-Fisher idealized population model, the conditional variance of the allele frequency p', given the allele frequency p in the previous generation, is

$$\operatorname{var}(p' \mid p) = \frac{p(1-p)}{2N}$$

Let $\widehat{\operatorname{var}}(p' \mid p)$ denote the same, typically larger, variance in the actual population under consideration. The variance effective population size $N_e^{(v)}$ is defined as the size of an idealized population with the same variance. This is found by substituting $\widehat{\operatorname{var}}(p' \mid p)$. for $\operatorname{var}(p' \mid p)$ and solving for N which gives

$$N_e^{(v)} = \frac{p(1-p)}{2\widehat{\operatorname{var}}(p)}$$

Theoretical Examples

In the following examples, one or more of the assumptions of a strictly idealised population are relaxed, while other assumptions are retained. The variance effective population size of the more relaxed population model is then calculated with respect to the strict model.

Variations in Population Size

Population size varies over time. Suppose there are t non-overlapping generations, then effective population size is given by the harmonic mean of the population sizes:

$$\frac{1}{N_e} = \frac{1}{t} \sum_{i=1}^{t} \frac{1}{N_i}$$

For example, say the population size was N = 10, 100, 50, 80, 20, 500 for six generations (t = 6). Then the effective population size is the harmonic mean of these, giving:

$\dfrac{1}{N_e}$	$= \dfrac{\dfrac{1}{10} + \dfrac{1}{100} + \dfrac{1}{50} + \dfrac{1}{80} + \dfrac{1}{20} + \dfrac{1}{500}}{6}$
	$= \dfrac{0.1945}{6}$
	$= 0.032416667$
N_e	$= 30.8$

Note this is less than the arithmetic mean of the population size, which in this example is 126.7. The harmonic mean tends to be dominated by the smallest bottleneck that the population goes through.

Dioeciousness

If a population is dioecious, i.e. there is no self-fertilisation then

$$N_e = N + \frac{1}{2}$$

or more generally,

$$N_e = N + \frac{D}{2}$$

where D represents dioeciousness and may take the value 0 (for not dioecious) or 1 for dioecious.

When N is large, N_e approximately equals N, so this is usually trivial and often ignored:

$$N_e = N + \frac{1}{2} \approx N$$

Variance in Reproductive Success

If population size is to remain constant, each individual must contribute on average two gametes to the next generation. An idealized population assumes that this follows a Poisson distribution so that the variance of the number of gametes contributed, k is equal to the mean number contributed, i.e. 2:

$$\text{var}(k) = \overline{k} = 2.$$

However, in natural populations the variance is often larger than this. The vast majority of individuals may have no offspring, and the next generation stems only from a small number of individuals, so

$$\text{var}(k) > 2.$$

The effective population size is then smaller, and given by:

$$N_e^{(v)} = \frac{4N - 2D}{2 + \text{var}(k)}$$

Note that if the variance of k is less than 2, N_e is greater than N. In the extreme case of a population experiencing no variation in family size, in a laboratory population in which the number of offspring is artificially controlled, $V_k = 0$ and $N_e = 2N$.

Non-Fisherian Sex-ratios

When the sex ratio of a population varies from the Fisherian 1:1 ratio, effective population size is given by:

$$N_e^{(v)} = N_e^{(F)} = \frac{4N_m N_f}{N_m + N_f}$$

Where N_m is the number of males and N_f the number of females. For example, with 80 males and 20 females (an absolute population size of 100):

N_e	$= \dfrac{4 \times 80 \times 20}{80 + 20}$
	$= \dfrac{6400}{100}$
	$= 64$

Again, this results in N_e being less than N.

Inbreeding Effective Size

Alternatively, the effective population size may be defined by noting how the average inbreeding coefficient changes from one generation to the next, and then defining N_e as the size of the idealized population that has the same change in average inbreeding coefficient as the population under consideration. The presentation follows Kempthorne (1957).

For the idealized population, the inbreeding coefficients follow the recurrence equation

$$F_t = \frac{1}{N}\left(\frac{1+F_{t-2}}{2}\right) + \left(1 - \frac{1}{N}\right)F_{t-1}.$$

Using Panmictic Index (1 – F) instead of inbreeding coefficient, we get the approximate recurrence equation

$$1 - F_t = P_t = P_0\left(1 - \frac{1}{2N}\right)^t.$$

The difference per generation is

$$\frac{P_{t+1}}{P_t} = 1 - \frac{1}{2N}.$$

The inbreeding effective size can be found by solving

$$\frac{P_{t+1}}{P_t} = 1 - \frac{1}{2N_e^{(F)}}.$$

This is

$$N_e^{(F)} = \frac{1}{2\left(1 - \dfrac{P_{t+1}}{P_t}\right)}$$

although researchers rarely use this equation directly.

Theoretical Example: Overlapping Generations and Age-structured Populations

When organisms live longer than one breeding season, effective population sizes have to take into account the life tables for the species.

Haploid

Assume a haploid population with discrete age structure. An example might be an organism that can survive several discrete breeding seasons. Further, define the following age structure characteristics:

v_i = Fisher's reproductive value for age i,

ℓ_i = The chance an individual will survive to age i, and

N_0 = The number of newborn individuals per breeding season.

The generation time is calculated as

$$T = \sum_{i=0}^{\infty} \ell_i v_i = \text{average age of a reproducing individual}$$

Then, the inbreeding effective population size is

$$N_e^{(F)} = \frac{N_0 T}{1 + \sum_i \ell_{i+1}^2 v_{i+1}^2 \left(\dfrac{1}{\ell_{i+1}} - \dfrac{1}{\ell_i}\right)}.$$

Diploid

Similarly, the inbreeding effective number can be calculated for a diploid population with discrete age structure. This was first given by Johnson, but the notation more closely resembles Emigh and Pollak.

Assume the same basic parameters for the life table as given for the haploid case, but distinguishing between male and female, such as N_0^f and N_0^m for the number of newborn females and males, respectively (notice lower case f for females, compared to upper case F for inbreeding).

The inbreeding effective number is

$$\frac{1}{N_e^{(F)}} = \frac{1}{4T} \left\{ \frac{1}{N_0^f} + \frac{1}{N_0^m} + \sum_i \left(\ell_{i+1}^f\right)^2 \left(v_{i+1}^f\right)^2 \left(\frac{1}{\ell_{i+1}^f} - \frac{1}{\ell_i^f}\right) \right.$$

$$\left. + \sum_i \left(\ell_{i+1}^m\right)^2 \left(v_{i+1}^m\right)^2 \left(\frac{1}{\ell_{i+1}^m} - \frac{1}{\ell_i^m}\right) \right\}.$$

Coalescent Effective Size

According to the neutral theory of molecular evolution, a neutral allele remains in a population for Ne generations, where Ne is the effective population size. An idealised diploid population will have a pairwise nucleotide diversity equal to $4\,\mu$ Ne, where μ is the mutation rate. The sojourn effective population size can therefore be estimated empirically by dividing the nucleotide diversity by the mutation rate.

The coalescent effective size may have little relationship to the number of individuals physically present in a population. Measured coalescent effective population sizes vary between genes in the same population, being low in genome areas of low recombination and high in genome areas of high recombination. Sojourn times are proportional to N in neutral theory, but for alleles under selection, sojourn times are proportional to log(N). Genetic hitchhiking can cause neutral mutations to have sojourn times proportional to log(N): this may explain the relationship between measured effective population size and the local recombination rate.

Selection Effective Size

In an idealised Wright-Fisher model, the fate of an allele, beginning at an intermediate frequency, is largely determined by selection if the selection coefficient s $\gg$ 1/N, and largely determined by neutral genetic drift if s $\ll$ 1/N. In real populations, the cutoff value of s may depend instead on local recombination rates. This limit to selection in a real population may be captured in a toy Wright-Fisher simulation through the appropriate choice of Ne. Populations with different selection effective population sizes are predicted to evolve profoundly different genome architectures.

Factors that Affect N_e

In general, N_e is less than the census population size N (the actual number of animals present). In some cases it is far lower. We will examine five forces acting to make N_e (effective population size) differ from N (census population size, or number of animals actually present).

1) One of the most important influences reducing N_e relative to N is fluctuating population size. This is because N_e that accounts for fluctuating population size is calculated as the harmonic mean of the census size. The harmonic mean is the reciprocal of [the average of the reciprocals]. An example will clarify both the meaning of this and its dramatic impact. Say we have a population census over four time periods of 200, 150, 50, and 300. What is the estimate of N_e?

$$N_e = \frac{1}{\left(\dfrac{1}{250} + \dfrac{1}{250} + \dfrac{1}{50} + \dfrac{1}{300}\right)/4} = 123.7$$

The harmonic mean of 123.7 contrasts with the arithmetic mean of 200. Another relevant paper is by Vucetich et al. (1997).

2) A second factor affecting N_e is the breeding sex ratio. A famous equation for dairy cows shows the dramatic effect a very skewed sex ratio can have. Say we have 96 cows and 4 bulls as the "breeding herd". What is N_e? The equation is:

$$N_e = \frac{4 \times N_m \times N_f}{N_m + N_f}$$

The sex ratio effective size of 15.4 is much closer to the number of bulls than to the number of cows. Try putting 50:50 in Eqn 7. 5 to verify that $N_e = N$ when the sex ratio is 1:1. [The sex ratio effect I have described is a special case of a more general impact on effective size due to variance in reproductive success].

3) A third influence on N_e can have an interesting effect that sometimes enters into captive breeding designs. N_e assumes a Poisson distribution of family (offspring) numbers. The Poisson is characterized by having the variance equal the mean. If the variance is lower than the mean, then N_e can actually be larger than the census size! Zoos will sometimes maintain their captive breeding stock to equalize family sizes (zero variance). This can increase N_e. They will usually need to keep "reserve" breeders, in case of the death of one of the selected breeders. In natural populations, if the environment causes the variance to exceed the mean (which may occur fairly frequently) then N_e will again be less than N. Hartl (2000, p. 96) gives an example of the reduction of N_e relative to N because of variance in family size.

4) Overlapping generations can also act to reduce N_e. (Felsenstein, 1971).

5) Yet another factor affecting N_e is the spatial dispersion (pattern of spatial distribution) of the population. Its influence on the effective size is given by:

$$N_e = 4\pi\sigma^2\delta$$

where σ^2 is the variance of the dispersal distance and δ is the density of individuals. This formulation is often called the neighborhood size and assumes a normal (bell-shaped) distribution of dispersal distances (out in a circular shape from the source, hence the π). So, again, changes in the variance of dispersal size can affect N_e (viscous populations will have smaller N_e). Note that the units of density are units of area (e.g., hectares) and that the variance will be distance squared (= area), so the distance units will cancel, giving us a "pure number" of individuals, which is what we should expect. Here's a place where good natural history and demographic data might produce reasonably good estimates of N_e. Hartl (2000, p. 98) gives an example of calculating N_e using this approach. Woolfenden and Fitzpatrick (1984) used this approach to estimate N_e for Florida Scrub-Jays -- their estimate of N_e was 298.

Population Growth

Population growth refers to change in the size of a population—which can be either positive or negative—over time, depending on the balance of births and deaths. If there are many deaths, the world's population will grow very slowly or can even decline. Population growth is measured in both absolute and relative terms. Absolute growth is the difference in numbers between a population over time; for example, in 1950 the world's population was 4 billion, and in 2000 it was 6 billion, a growth of 2 billion. Relative growth is usually expressed as a rate or a percentage; for example, in 2000 the rate of global population growth was 1.4 percent (or 14 per 1,000). For every 1,000 people in the world, 14 more are being added per year.

For the world as a whole, population grows to the extent that the number or rate of births exceeds the number or rate of deaths. The difference between these numbers (or rates) is termed "natural increase" (or "natural decrease" if deaths exceed births). For example, in 2000 there were 22 births per 1,000 population (the number of births per 1,000 population is termed the "crude birth rate") and 9 deaths per 1,000 population (the "crude death rate"). This difference accounts for the 2000 population growth rate of 14 per 1,000, which is also the rate of natural increase. In absolute numbers, this means that approximately 78 million people—or about the population of the Philippines—are added to the world each year. For countries, regions, states, and so on, population growth results from a combination of natural increase and migration flows. The rate of natural increase is equivalent to the rate of population growth only for the world as a whole and for any smaller geographical units that experience no migration.

Populations can grow at an exponential rate, just as compound interest accumulates in a bank account. One way to assess the growth potential of a population is to calculate its doubling time—the number of years it will take for a population to double in size, assuming the current rate of population growth remains unchanged. This is done by applying the "rule of seventy"; that is, seventy divided by the current population growth rate (in percent per year). The 1.4 percent global population growth rate in 2000 translates into a doubling time (if the growth rate remains constant) of fifty-one years.

Population growth did not become exponential until around 1750. Before that, high mortality counterbalanced the high fertility needed by agrarian parents. Death rates were high and life expectancy was low; life expectancy at birth was in the range of twenty to forty years (most likely around thirty years) until the middle of the eighteenth century. This high mortality was a function of several factors, including poor nutrition, which led directly to deaths through starvation and indirectly through increasing susceptibility to disease; epidemics; and, quite possibly, infanticide and geronticide, especially during times of food shortage.

Starting in the middle of the eighteenth century, the mortality rate began to decline in the West, the first place in the world where the natural balance between births and deaths was altered by humans. This decline in deaths occurred not because of major medical breakthroughs (e.g., penicillin was first used only in the 1940s) but rather because of improvements in food availability, housing, water cleanliness, personal hygiene, and public sanitation. Later, in the twentieth century, medical advances, particularly vaccinations against infectious diseases, accelerated mortality decline. Western mortality decline was relatively slow, paralleling socioeconomic development, and it occurred in a global context in which European population "surplus" (arising from gaps between lowering mortality and more slowly lowering fertility) was able to migrate to new areas (e.g., the United States, Canada, and Australia) that were very sparsely populated by Aboriginal peoples (whose numbers were reduced even more by contagious diseases brought by Europeans).

The "population growth rate" is the rate at which the number of individuals in a population increases in a given time period, expressed as a fraction of the initial population. Specifically, population growth rate refers to the change in population over a unit time period, often expressed as a percentage of the number of individuals in the population at the beginning of that period. This can be written as the formula, valid for a sufficiently small time interval:

$$\text{Population growth rate} = \frac{P(t_2) - P(t_1)}{P(t_1)(t_2 - t_1)}$$

A positive growth rate indicates that the population is increasing, while a negative growth rate indicates that the population is decreasing. A growth ratio of zero indicates that there were the same number of individuals at the beginning and end of the period—a growth rate may be zero even when there are significant changes in the birth rates, death rates, immigration rates, and age distribution between the two times.

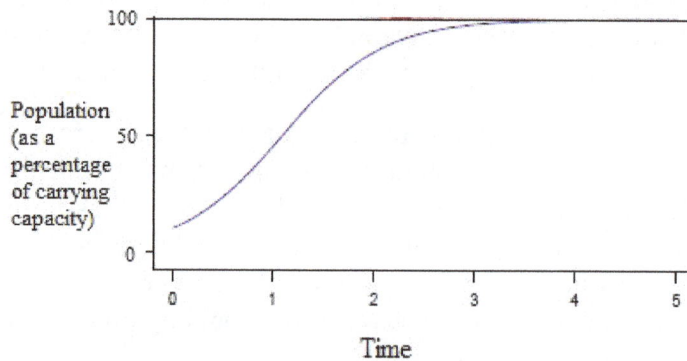

The logistic growth of a population.

A related measure is the net reproduction rate. In the absence of migration, a net reproduction rate of more than 1 indicates that the population of females is increasing, while a net reproduction rate less than one (sub-replacement fertility) indicates that the population of females is decreasing.

Most populations do not grow exponentially, rather they follow a logistic model. Once the population has reached its carrying capacity, it will stabilize and the exponential curve will level off towards the carrying capacity, which is usually when a population has depleted most its natural resources.

Logistic Equation

The growth of a population can often be modelled by the logistic equation

$$\frac{dP}{dt} = kP\left(1 - \frac{P}{K}\right),$$

where

- $P(t)$= the population after time t;
- t= time a population grows;
- k= the relative growth rate coefficient;

- $K=$ the carrying capacity of the population; defined by ecologists as the maximum population size that a particular environment can sustain.

As it is a separable differential equation, the population may be solved explicitly, producing a logistic function:

$$P(t) = \frac{K}{1 + Ae^{-rt}},$$

where $A = \frac{K - P_0}{P_0}$ and P_0 is the initial population at time 0.

Human Population Growth Rate

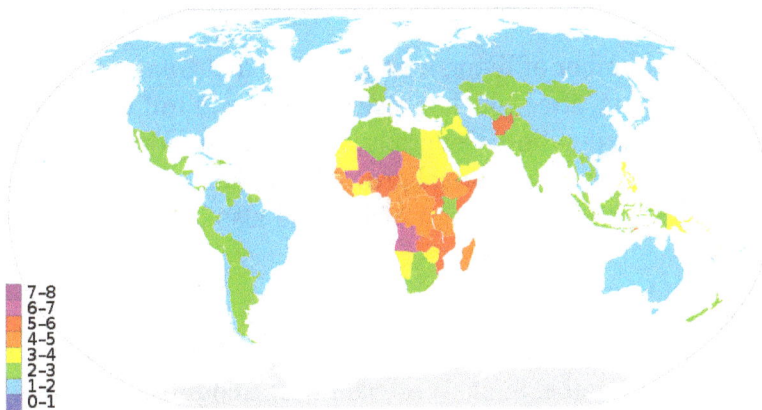

A world map showing global variations in fertility rate per woman

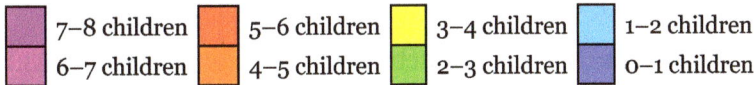

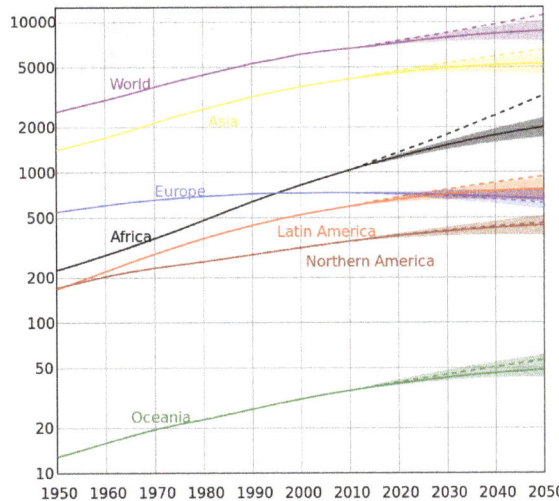

7–8 children	5–6 children	3–4 children	1–2 children
6–7 children	4–5 children	2–3 children	0–1 children

Estimates of population evolution in different continents between 1950 and 2050 according to the United Nations. The vertical axis is logarithmic and is in millions of people.

In 2017, the estimated annual growth rate was 1.1%. The CIA World Factbook gives the world annual birthrate, mortality rate, and growth rate as 1.86%, 0.78%, and 1.08% respectively. The last 100 years have seen a massive fourfold increase in the population, due to medical advances, lower mortality rates, and an increase in agricultural productivity made possible by the Green Revolution.

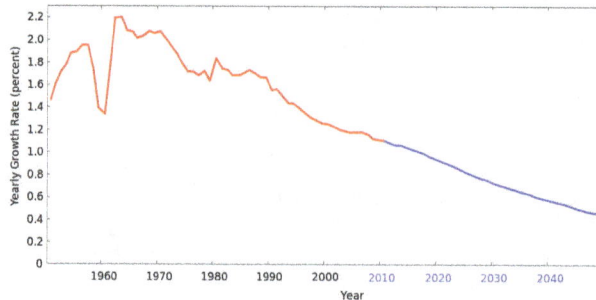

World population growth rates between 1950–2050

The annual increase in the number of living humans peaked at 88.0 million in 1989, then slowly declined to 73.9 million in 2003, after which it rose again to 75.2 million in 2006. In 2017, the human population increased by 83 million. Generally, developed nations have seen a decline in their growth rates in recent decades, though annual growth rates remain above 2% in poverty-stricken countries of the Middle East and Sub-Saharan Africa, and also in South Asia, Southeast Asia, and Latin America.

In some countries the population is declining, especially in Eastern Europe, mainly due to low fertility rates, high death rates and emigration. In Southern Africa, growth is slowing due to the high number of AIDS-related deaths. Some Western Europe countries might also experience population decline. Japan's population began decreasing in 2005; it now has the highest standard of living in the world.

The United Nations Population Division projects world population to reach 11.2 billion by the end of the 21st century, but Sanjeev Sanyal has argued that global fertility will fall below the replacement rate in the 2020s and that world population will peak below 9 billion by 2050, followed by a long decline. A 2014 study in Science concludes that the global population will reach 11 billion by 2100, with a 70% chance of continued growth into the 22nd century.

Growth by Country

According to United Nations population statistics, the world population grew by 30%, or 1.6 billion humans, between 1990 and 2010. In number of people the increase was highest in India (350 million) and China (196 million). Population growth was among highest in the United Arab Emirates (315%) and Qatar (271%).

Growth rates of the world's most populous countries				
Rank	Country	Population 2010	Population 1990	Growth (%) 1990–2010
	World	6,895,889,000	5,306,425,000	30.0%
1	China	1,341,335,000	1,145,195,000	17.1%
2	India	1,224,614,000	873,785,000	40.2%
3	United States	310,384,000	253,339,000	22.5%

4	Indonesia	239,871,000	184,346,000	30.1%
5	Brazil	194,946,000	149,650,000	30.3%
6	Pakistan	173,593,000	111,845,000	55.3%
7	Nigeria	158,423,000	97,552,000	62.4%
8	Bangladesh	148,692,000	105,256,000	41.3%
9	Russia	142,958,000	148,244,000	-3.6%
10	Japan	128,057,000	122,251,000	4.7%

Growth by Region

Population growth rates vary by world region, with the highest growth rates in Sub-Saharan Africa and the lowest in Europe. For example, from 1950 to 2010, Sub-Saharan African grew over three and a half times, from about 186 million to 856 million. On the other hand, Europe only increased by 35%, from 547 million in 1950 to 738 million in 2010. As a result of these varying population growths, Sub-Saharan Africa changed from 7.4% of world population in 1950 to 12.4% in 2010, while Europe declined from 22% to 11% in the same time period.

Into the Future

According to the UN's 2017 revision to its population projections, world population is projected to reach 11.2 billion by 2100 compared to 7.6 billion in 2017. In 2011, Indian economist Sanjeev Sanyal disputed the UN's figures and argued that birth rates will fall below replacement rates in the 2020s. According to his projections, population growth will be only sustained till the 2040s by rising longevity, but will peak below 9 bn by 2050. Conversely, a 2014 paper by demographers from several universities and the United Nations Population Division projected that the world's population would reach about 10.9 billion in 2100 and continue growing thereafter. One of its authors, Adrian Raftery, a University of Washington professor of statistics and of sociology, says "The consensus over the past 20 years or so was that world population, which is currently around 7 billion, would go up to 9 billion and level off or probably decline. We found there's a 70 percent probability the world population will not stabilize this century. Population, which had sort of fallen off the world's agenda, remains a very important issue."

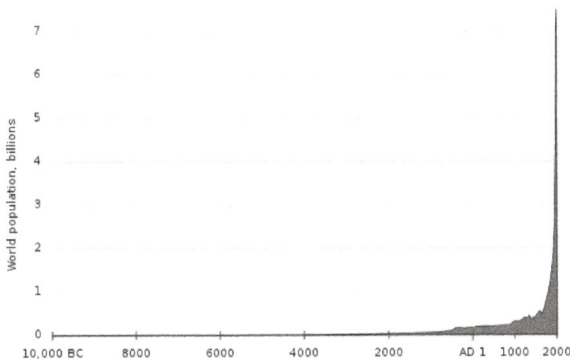

Estimated size of human population from 10,000 BCE to 2000 CE.

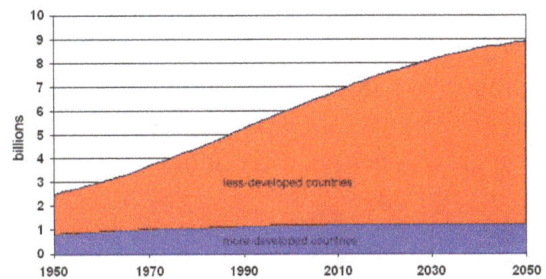

Population Growth in More- and Less-Developed Countries, 2002.

Source: United Nations, World Population Prospects.

The majority of world population growth today is occurring in less developed countries.

Theories of Population Growth

While theories about population growth first appeared in ancient Greece, the English clergyman and economist Thomas Malthus (1766–1834) is considered to be the pioneering theorist of the modern age. Malthus formulated a "principle of population" that held that unchecked population grows more quickly than the means of subsistence (food and resources) to sustain it. Population will be controlled either by preventive checks (lowering the number of births, particularly by postponement of marriage age) or by positive checks (increasing deaths as a result of famines, plagues, natural disasters, war). Given a morally based preference for preventive checks, later followers of Malthus (neo-Malthusians) have supported family planning and contraception even though Malthus himself felt that contraception was unacceptable. Other neo-Malthusians have focused upon the claimed negative effects of rapid population growth: war, violence, and environmental degradation.

Karl Marx's views on population were directly opposed to those of Malthus. Marx disagreed with the Malthusian idea of a universal principle of population that applied to all societies. For Marx, population growth depended upon the economic base of society. Thus, capitalist society is characterized by its own population principle, which Marx termed the "law of relative population surplus." He argued that capitalism creates overpopulation (i.e., a surplus of people relative to jobs), leading to increased unemployment, cheap labor, and poverty. Also, capitalism requires unemployment in order to ensure a docile, low-paid class of laborers. Marx envisioned that overpopulation would not occur in postcapitalist, communist society.

In the middle of the twentieth century, demographic transition theory became the dominant theory of population growth. Based on observed trends in Western European societies, it argues that populations go through three stages in their transition to a modern pattern. Stage One (pretransition) is characterized by low or no growth, and high fertility is counterbalanced by high mortality. In Stage Two (the stage of transition), mortality rates begin to decline, and the population grows at a rapid pace. By the end of this stage, fertility has begun to decline as well. However, because mortality decline had a head start, the death rate remains lower than the birth rate, and the population continues to experience a high rate of growth. In Stage Three (posttransition), the movement to low fertility and mortality rates is complete, producing once again a no-growth situation. The theory of demographic transition explains these three stages in terms of economic development, namely industrialization and urbanization. Since about 1980, demographic transition theory has been criticized on a number of grounds, including its assumption that the demographic experience of non-Western societies will inevitably follow that of the West; its failure to consider cultural variables; and its hypothesized relationship between population growth and economic development. Indeed, all three theories above contain assumptions about population growth and economic development; however, there is mounting evidence that this relationship is complex and varies from context to context. As the twenty-first century begins, the attempt to erect a general theory of population growth has been abandoned, signaling for some an alarming trend in population studies.

Population Dynamics

A population describes a group of individuals of the same species occupying a specific area at a specific time. Some characteristics of populations that are of interest to biologists include the population density , the birthrate , and the death rate . If there is immigration into the population, or emigration out of it, then the immigration rate and emigration rate are also of interest. Together, these population parameters, or characteristics, describe how the population density changes over time. The ways in which population densities fluctuate—increasing, decreasing, or both over time—is the subject of population dynamics.

Population density measures the number of individuals per unit area, for example, the number of deer per square kilometer. Although this is straightforward in theory, determining population densities for many species can be challenging in practice.

Measuring Population Density

One way to measure population density is simply to count all the individuals. This, however, can be laborious. Alternatively, good estimates of population density can often be obtained via the quadrat method. In the quadrat method, all the individuals of a given species are counted in some subplot of the total area. Then that data is used to figure out what the total number of individuals across the entire habitat should be.

The quadrat method is particularly suited to measuring the population densities of species that are fairly uniformly distributed over the habitat. For example, it has been used to determine the population density of soil species such as nematode worms. It is also commonly used to measure the population density of plants.

For more mobile organisms, the capture-recapture method may be used. With this technique, a number of individuals are captured, marked, and released. After some time has passed, enough time to allow for the mixing of the population, a second set of individuals is captured. The total population size may be estimated by looking at the proportion of individuals in the second capture set that are marked. Obviously, this method works only if one can expect individuals in the population to move around a lot and to mix. It would not work, for example, in territorial species, where individuals tend to remain near their territories.

The birthrate of a population describes the number of new individuals produced in that population per unit time. The death rate, also called mortality rate, describes the number of individuals who die in a population per unit time. The immigration rate is the number of individuals who move into a population from a different area per unit time. The emigration rates describe the numbers of individuals who migrate out of the population per unit time.

The values of these four population parameters allow us to determine whether a population will increase or decrease in size. The "intrinsic rate of increase r " of a population is defined as r = (birth rate immigration rate +)-(death rate + emigration rate).

If r is positive, then more individuals will be added to the population than lost from it. Consequently, the population will increase in size. If r is negative, more individuals will be lost from the population than are being added to it, so the population will decrease in size. If r is exactly zero, then the population size is stable and does not change. A population whose density is not changing is said to be at equilibrium .

Intrinsic Rate of Increase

The rate at which a population increases in size if there are no density-dependent forces regulating the population is known as the intrinsic rate of increase. It is

$$\frac{dN}{dt}\frac{1}{N} = r$$

where the derivative dN/dt is the rate of increase of the population, N is the population size, and r is the intrinsic rate of increase. Thus r is the maximum theoretical rate of increase of a population per individual – that is, the maximum population growth rate. The concept is commonly used in insect population biology to determine how environmental factors affect the rate at which pest populations increase.

Common Mathematical Models

Exponential Population Growth

Exponential growth describes unregulated reproduction. It is very unusual to see this in nature. In the last 100 years, human population growth has appeared to be exponential. In the long run, however, it is not likely. Paul Ehrlich and Thomas Malthus believed that human population growth would lead to overpopulation and starvation due to scarcity of resources. They believed that human population would grow at rate in which they exceed the ability at which humans can find food. In the future, humans would be unable to feed large populations. The biological assumptions of exponential growth is that the per capita growth rate is constant. Growth is not limited by resource scarcity or predation.

Simple Discrete Time Exponential Model

$$N_{t+1} = \lambda N_t$$

where λ is the discrete-time per capita growth rate. At $\lambda = 1$, we get a linear line and a discrete-time per capita growth rate of zero. At $\lambda < 1$, we get a decrease in per capita growth rate. At $\lambda > 1$, we get an increase in per capita growth rate. At $\lambda = 0$, we get extinction of the species.

Continuous Time Version of Exponential Growth

Some species have continuous reproduction.

$$\frac{dN}{dT} = rN$$

where $\frac{dN}{dT}$ is the rate of population growth per unit time, r is the maximum per capita growth rate, and N is the population size.

At $r > 0$, there is an increase in per capita growth rate. At $r = 0$, the per capita growth rate is zero. At $r < 0$, there is a decrease in per capita growth rate.

Logistic Population Growth

"Logistics" comes from the French word logistique, which means "to compute". Population regulation is a density-dependent process, meaning that population growth rates are regulated by the density of a population. Consider an analogy with a thermostat. When the temperature is too hot, the thermostat turns on the air conditioning to decrease the temperature back to homeostasis. When the temperature is too cold, the thermostat turns on the heater to increase the temperature back to homeostasis. Likewise with density dependence, whether the population density is high or low, population dynamics returns the population density to homeostasis. Homeostasis is the set point, or carrying capacity, defined as K.

Continuous-time Model of Logistic Growth

$$\frac{dN}{dT} = rN\left(1 - \frac{N}{K}\right)$$

where $\left(1 - \frac{N}{K}\right)$ is the density dependence, N is the number in the population, K is the set point for homeostasis and the carrying capacity. In this logistic model, population growth rate is highest at 1/2 K and the population growth rate is zero around K. The optimum harvesting rate is a close rate to 1/2 K where population will grow the fastest. Above K, the population growth rate is negative. The logistic models also show density dependence, meaning the per capita population growth rates decline as the population density increases. In the wild, you can't get these patterns to emerge without simplification. Negative density dependence allows for a population that overshoots the carrying capacity to decrease back to the carrying capacity, K.

According to R/K selection theory organisms may be specialised for rapid growth, or stability closer to carrying capacity.

Discrete time Logistical Model

$$N_{t+1} = N_t + rN_t(1 - N_t / K)$$

This equation uses r instead of λ because per capita growth rate is zero when r = 0. As r gets very high, there are oscillations and deterministic chaos. Deterministic chaos is large changes in population dynamics when there is a very small change in r. This makes it hard to make predictions at high r values because a very small r error results in a massive error in population dynamics.

Population is always density dependent. Even a severe density independent event cannot regulate populate, although it may cause it to go extinct.

Not all population models are necessarily negative density dependent. The Allee effect allows for a positive correlation between population density and per capita growth rate in communities with very small populations. For example, a fish swimming on its own is more likely to be eaten than the same fish swimming among a school of fish, because the pattern of movement of the school of fish is more likely to confuse and stun the predator.

Individual-based Models

Cellular automata are used to investigate mechanisms of population dynamics. Here are relatively simple models with one and two species.

Logical deterministic individual-based cellular automata model of single species population growth

Logical deterministic individual-based cellular automata model of interspecific competition for a single limited resource

Fisheries and Wildlife Management

In fisheries and wildlife management, population is affected by three dynamic rate functions.

- Natality or birth rate, often recruitment, which means reaching a certain size or reproductive stage. Usually refers to the age a fish can be caught and counted in nets.

- Population growth rate, which measures the growth of individuals in size and length. More important in fisheries, where population is often measured in biomass.

- Mortality, which includes harvest mortality and natural mortality. Natural mortality includes non-human predation, disease and old age.

If N_1 is the number of individuals at time 1 then

$$N_1 = N_0 + B - D + I - E$$

where N_0 is the number of individuals at time 0, B is the number of individuals born, D the number that died, I the number that immigrated, and E the number that emigrated between time 0 and time 1.

If we measure these rates over many time intervals, we can determine how a population's density changes over time. Immigration and emigration are present, but are usually not measured.

All of these are measured to determine the harvestable surplus, which is the number of individuals that can be harvested from a population without affecting long-term population stability or average population size. The harvest within the harvestable surplus is termed "compensatory" mortality, where the harvest deaths are substituted for the deaths that would have occurred naturally. Harvest above that level is termed "additive" mortality, because it adds to the number of deaths that would have occurred naturally. These terms are not necessarily judged as "good" and "bad," respectively, in population management. For example, a fish & game agency might aim to reduce the size of a deer population through additive mortality. Bucks might be targeted to increase buck competition, or does might be targeted to reduce reproduction and thus overall population size.

For the management of many fish and other wildlife populations, the goal is often to achieve the largest possible long-run sustainable harvest, also known as maximum sustainable yield (or MSY). Given a population dynamic model, such as any of the ones above, it is possible to calculate the population size that produces the largest harvestable surplus at equilibrium. While the use of population dynamic models along with statistics and optimization to set harvest limits for fish and game is controversial among scientists, it has been shown to be more effective than the use of human judgment in computer experiments where both incorrect models and natural resource management students competed to maximize yield in two hypothetical fisheries. To give an example of a non-intuitive result, fisheries produce more fish when there is a nearby refuge from human predation in the form of a nature reserve, resulting in higher catches than if the whole area was open to fishing.

For Control Applications

Population dynamics have been widely used in several control theory applications. With the use of evolutionary game theory, population games are broadly implemented for different industrial and daily-life contexts. Mostly used in multiple-input-multiple-output (MIMO) systems, although they can be adapted for use in single-input-single-output (SISO) systems. Some examples of applications are military campaigns, resource allocation for water distribution, dispatch of distributed generators, lab experiments, transport problems, communication problems, among others. Furthermore, with the adequate contextualization of industrial problems, population dynamics can be an efficient and easy-to-implement solution for control-related problems. Multiple academic research has been and is continuously carried out.

Abundance

Species diversity is determined not only by the number of species within a biological community—i.e., species richness—but also by the relative abundance of individuals in that community. Species abundance is the number of individuals per species, and relative abundance refers to the evenness of distribution of individuals among species in a community. Two communities may be equally rich in species but differ in relative abundance. For example, each community may contain 5 species and 300 individuals, but in one community all species are equally common (e.g., 60 individuals of each species), while in the second community one species significantly outnumbers the other four.

These components of species diversity respond differently to various environmental conditions. A region that does not have a wide variety of habitats usually is species-poor; however, the few species that are able to occupy the region may be abundant because competition with other species for resources will be reduced.

ACFOR Scale

The ACFOR scale can be used to collect data on the abundance of each species

- Abundant
- Common
- Frequent
- Occasional
- Rare
- None

Data collected on the ACFOR scale is often biased or subjective. Many investigators over-estimate conspicuous plants (especially those in flower) and under-estimate inconspicuous plants. More reliable data will come from quantitative methods.

r/K Selection Theory

The science behind r/K Selection theory was hashed out decades ago. It emerged as biologists pondered why some species reproduced slowly using monogamy and high-investment parenting, while other species reproduced explosively, using promiscuity and single parenting. At the time this science was developed, the researchers were wholly unaware of its relevance to our modern ideological battles in the world of politics. The terms r and K came from variables in equations which described how populations would change over time. r represented the maximal reproductive rate of an individual, while K represented the carrying capacity of an environment.

r/K selection theory describes two environmental extremes, and the strategies a population will produce to exploit each extreme. As a result of these strategies, each of these two environments will produce a very particular psychology in the individuals exposed to them.

The first environment an organism may face is the presence of freely available resources, which is referred to as an r-selective environment. This most often occurs when a predator keeps a population consistently lower than the carrying capacity of its environment. Just as rabbits do not strip their grassy fields bare due to the predation they endure, the r-strategy is designed to exploit an environment where resources are freely available, everywhere.

In r-selection, those individuals who waste time fighting for food will be out-reproduced by pacifists, who simply focus upon eating, and reproducing. Fighting also entails risks of injury or death – risks which are pointless given the free availability of resources everywhere. Hence this environment will favor a tendency towards conflict avoidance, and tend to cull the aggressive and competitive. It will also evolve tendencies towards mating as early as possible, as often as possible, with as many mates as possible, while investing as little effort as possible rearing offspring. Here, there are unlimited resources just waiting to be utilized, and even the most unfit can acquire them. As a result, it is more advantageous to produce as many offspring as possible, as quickly as possible, regardless of fitness, so as to out-reproduce those who either waste time producing quality offspring or waste time competing with each other.

Since group competition will not arise in the r-selected environment, r-type organisms will not exhibit loyalty to fellow members of their species, or a drive to sacrifice on their behalf. Indeed, the very notion of in-group will be foreign, and the concept of personal sacrifice for other in-group members will be wholly alien. This is why rabbits, mice, antelope, and other r-selected species, although pleasant, will tend to not exhibit any loyalty or emotional attachment to peers. When resources are freely available, group competition is a risk one need not engage in to acquire resources, so this loyalty to in-group and emotional attachment to peers is not favored.

Here in the r-strategy, we see the origins of the Liberal's tendencies towards conflict avoidance, from oppositions to free-market capitalism, to pacifism, to demands that all citizens disarm so as to avoid any chance of conflict and competition. Even the newer tendencies to support the "everyone gets a trophy" movement are outgrowths of this competition-averse urge, and desire for free resource availability. Similarly, Liberals are supportive of promiscuity, supportive of efforts to expose children to ever earlier sexual education, and, as the debate over Murphy Brown showed, Liberals are supportive of low-investment, single parenting. Finally, as John Jost has shown, Liberals show diminished loyalty to in-group, similar to how r-selected organisms do not fully understand the reason for even perceiving an in-group in nature.

In the other environment, a population exists at the carrying capacity of its environment. Since there is not enough food to go around, and someone must die from starvation, this will evolve a specific psychology within such a species.

Termed a K-type psychology, or K-Selected Reproductive Strategy, this psychology will embrace competitions between individuals and accept disparities in competitive outcomes as an innate part of the world, that is not to be challenged. Since individuals who do not fight for some portion of the limited resources will starve, this environment will favor an innately competitive, conflict-prone psychology. Study shows, such a psychology will also tend to embrace monogamy, embrace chastity until monogamous adulthood, and favor high-investment, two-parent parenting, with an emphasis upon rearing as successful an offspring as possible. This sexual selectiveness, mate monopolization, and high-investment rearing is all a form of competing to produce fitter offspring than peers. This evolves, because if one's offspring are fitter than the offspring of peers, they will be likely to acquire resources themselves, and reproduce successfully.

Although total numbers of offspring will be diminished with this high-investment rearing strategy, the offspring's success in competition is what is most important in a K-selective environment. Here, wasting time producing numerous offspring that are not as fit as possible will doom one to Darwinian failure. As time goes on, and K-selection continues, forming into competitive groups will often emerge as a strategy to acquire resources. This will add add loyalty to in-group to the suite of K-type psychological characteristics. This is why when we look at K-selected species in nature, we see packs of wolves, herds of elephants, prides of lions, and pods of dolphins, and each individual is loyal to their group and its competitive success. Since the only way to survive will be to acquire one's resources by out-competing peers, this invariably produces tremendously fast rates of evolutionary advancement. For this reason, K-selected organisms are usually more evolutionarily advanced than their r-selected counterparts, and will exhibit more complex adaptations, from increased intelligence and sentience, to increased physical capabilities, to loyalty and prosociality, in species where group competition occurs.

Clearly, this mirrors the Conservative's embrace of competitions, such as war, capitalism, and even the bearing of arms in self-defense against criminals. It also mirrors the Conservatives tendency to favor family values, such as abstinence until monogamy and two-parent parenting. It even explains why Conservatives feel driven to see their nation succeed as greatly as possible, regardless of the effects this has upon other nations or just members of their out-group.

To my eye, it is inherently clear that this r/K divergence is the origin of our political divide. Indeed, while policy proposals from Conservatives are predicated upon the premise that resources are inherently limited, and individuals should have to work and demonstrate merit to acquire them, Liberals advocate on behalf of policy proposals which seem to be predicated upon an assumption that there are always more than sufficient resources to let everyone live lives of equal leisure. To a Liberal, any scarcity must clearly arise due to some individual's personal greed and evil altering a natural state of perpetual plenty.

Here, we see how these two deeply imbued psychologies generate grossly different perceptual frameworks within those who are imbued with them. Just as a Liberal will never grasp why a Conservative will look down upon frequent promiscuity and single parenting, the Conservative will never grasp why the Liberal will be so firmly opposed to free market Capitalism, or the right to self defense when threatened. Each sees an inherently different world, and is programmed to desire an inherently different environment.

In nature, since it is the individuals who best exemplify this r-selected psychological standard who will reproduce under conditions of resource abundance, their offspring will carry these traits. As time goes on, the population will gradually develop ever more extreme presentations of these traits. As we show, there is copious evidence that a genetic allele, which diminishes dopamine signaling, is associated with every facet of the r-strategy's psychology, as well as a predisposition towards political Liberalism.

In addition, the r-strategy may have evolved to be engendered within individuals by environmental stimuli as well, through a desensitization to the neurotransmitter dopamine. This effect arises from its copious release in such an environment down-regulating receptor expression and consequently reducing receptor densities in nervous tissue. We also maintain that a lack of adversity in the environment will fail to develop a drive or ability to confront adversity, through a failure to develop a brain structure called the amygdala. In summary, an organism placed in an environment devoid of adversity, and filled with pleasure, may find itself more demanding of pleasure and less tolerant of adversity, than an organism which is enured to a less hospitable environment.

Within r/K selection theory, all populations will contain some differing degrees of r and K selected psychologies. As an environment shifts to one extreme or the other, a population will adopt a more r or K-selected psychology, but this will only last as long as the environmental conditions which produced the shift continue. Under conditions of reduced mortality, and copious resource availability, both r and K-selected psychologies will be present. This will continue until such time as resources become limited, and a competitive, K-selected pressure takes hold, or predation begins to cull both sides evenly, and the K-selected individuals, being slower reproducers are relatively culled back.

Interestingly, r/K Theory not only explains a means by which our political ideologies are adaptive to a specific environment. Many have noted an increasingly masculine quality to the women in our culture, as well as a corresponding effeminate nature to our men. Rush Limbaugh will often refer to them as the Feminazis, and the Castrati. In nature, a K-selected model of rearing involves a feminine mother, who nurtures offspring and guides them away from danger, combined with a more masculine male who will aggressively confront dangers, so as to protect his family.

However, when a population becomes increasingly r-selected, the nature of the sexual dimorphism and these sex-specific rearing behaviors will change. As you see a more r-strategy emerge, females of the species will need to become increasingly aggressive and masculine, since due to paternal abandonment, they must provision and protect their offspring alone. Since r-selected males are solely concerned with mating (before abandoning their mate), and fleeing from conflict, they become more diminutive, and more cowardly. The end result is the r-strategy has, inherent within it, a model of aggressive, manly females who raise children alone, and diminutive, effete males who are solely concerned with superficial, mate-attracting flash, and conflict avoidance.

Even more interestingly, as we point out in this blog post, as well as this blog post, there is evidence indicating that this phenomenon, accidentally over-expressed, may be responsible for producing males who are so effeminate that they are actually homosexual, and females who are so manly, they cross the boundary into lesbianism. Not only do the rearing behaviors and sexual characteristics change, but the males become attracted to more manly characteristics (which are now exhibited by the most adaptive females), and the females become more attracted to effeminate characteristics (which are now exhibited by the most adaptive males).

Some will ask, why would we have evolved both of these psychologies, within our species, instead of trending totally r or K. This can occur for a number of reasons. Obviously an organism which inhabits an environment where resources surge in availability, and then become scarce can see its r-types surge in number during times of plenty, only to die back once resources become scarce. Indeed, such a population may eventually see its individuals adapt to change their strategy with the availability of resources. Or, as time goes on, the r-types may evolve strategies designed to see a few members persist during times of scarcity, so they may explode again once resources become plentiful.

But in humans, the mechanism was probably a little more complex. When we first evolved, a critical adaptation was our loss of body hair. It allowed us to move about in the heat of an African day, when all other furred prey needed to bed down. To acquire meat, all we needed to do was roust a bedded down antelope, make it run a short distance, and it would rapidly collapse of heat stroke, so we could then acquire its meat. There are tribes in Africa who still hunt using this method.

This allowed us to explode in numbers, but as in all ecosystems, we eventually found there were not enough resources to support the population. It was at this time that our population divided.

At this point, the competition was fierce. One group adopted the K-selected psychology, stayed put, and slugged it out for resources, in free, merit based competition. They formed into groups, battled for territory and resources, and adopted a competitive, K-selected reproductive strategy. They became the K-type cohort of our population, embracing freedom and self-determination, free competition, monogamy, strong family values, loyalty to in-group, and sexual chastity in the youth.

As the battles began to rage, another cohort, more cowardly and weak, fled. Those who fled the fastest and the farthest, found themselves in a new, untapped territory, with free resource availability yet again. Those among them who did the best from Darwin's perspective, were those who adopted the most r-type strategy of free promiscuity, single parenting, and early age at first intercourse. They had no need for loyalty to in-group, and indeed, would have adopted a more selfish and cowardly psychology, to better disperse their genes, and serve their own self interests. They became our population's r-type cohort, and even today, the gene which is associated with Liberalism is found in large numbers in migratory populations, even as social psychologists note that Liberals score highly in novelty seeking, such as preferring new and novel environments, or unusual foods.

As time went on, Homo sapiens likely spread across the globe in this manner. r-types fled as the territory behind them became K-selective and competitive. As time went on, this constant selective pressure favoring fleeing gradually made the r-type more prone to flee competitions and adhere to an r-type mating strategy, and less able to even comprehend why K-types would ever seek monogamy or aggression when threatened, or innately perceive an in-group in need of defense.

In between where the r-types fled to, and where the K-types were battling it out, there was likely a sort of geographical spectrum. At one end were the extreme r-types on the frontier, and at the other were the extreme K-types, battling with neighbors. But in the middle, were areas where some r-types were mingling with some K-types. It is likely that there, these two strategies were evolving psychological traits which would allow them to persist in a mixed population. K-types tried to purge the disloyalty, selfishness, and promiscuity of the r-types, while r-types tried to use deception, as well as the rule-breaking and lack of loyalty identified by Jost (himself a Liberal), as an advantage. It would not surprise me if our political animus was evolved.

It is also interesting to note, even today, as r-types gain hold in a civilization, they seek to provide the unproductive and uncompetitive with the free resource availability of the r-selected environment. As in nature, as this goes on, the r-type cohort grows in the population, until the entire financial ecosystem collapses, the government dissolves, and the civilization becomes ruthlessly competitive. As in nature, free resource availability cannot go on forever.

To be clear, individuals are complex. Just as it is difficult to characterize a single individual organism's exact reproductive strategy in nature, it is difficult to characterize a single human's political strategy. However, just as the quantum mechanical world yields the chaos of its uncertainty to the order and formality of Newtonian physics when viewed from a distance, as we zoom out from our society we will find two primary ideologies within it. Just as in nature, these two ideologies match exactly the two psychologies of the r and K-type psychology.

Before closing, I would like to note that the primary environmental condition favoring an r-strategy is free resource availability. Too often the r-strategy is portrayed as a defensive adaptation designed solely to overcome the mortality of predation, or other forms of environmental harshness, through increased reproductive rates. The r-strategy however, is just as much an offensive adaptation designed to exploit free resource availability, and the absence of competitive selections for survival and reproduction.

In the book, we describe how this may be seen most clearly in the world of microbiology. There, complex, highly-adapted microbes are often drawn from a harsh, highly selective environment, and transferred to an unselective environment of ideal conditions and free resource availability (such as a petri dish of nutrient media housed in an incubator). There, they initially grow slowly, as each parent cell carefully produces colonies full of highly adapted daughter cells.

Some parent cells however, make mistakes, and produce less complex offspring, who reproduce more rapidly, as they devote less energy to their parent cell's complex adaptations. As time goes on a highly evolved isolate can quickly shed its adaptations and devolve into a strain of simpler, less complex cells which grow colonies astonishingly quickly on agar. Over time, if given only free resource availability, the cells of the simpler dysgenic strain will numerically dominate any peers which retain their complexity and adaptation. In this environment, due to the absence of competitive selections favoring fitness or complexity, the sole determinate of survival becomes sheer numerical advantage. As a result, it is this standard which the organism will evolve towards, and one will increasingly find a less complex, less evolved organism devoted solely to mating and reproduction. Free resource availability, and an absence of competitive selection pressure, by itself, is all that is necessary to fuel a rapid growth in the r-strategist cohort within a population.

In closing, it is impossible to deny that every aspect of political ideology revolves around the same fundamental issues of behavior that r/K selection theory revolves around. Although our species' embrace of group competition has further molded these urges, this is the evolutionary foundation of ideology. It is where political ideology began. For that reason, no individual can ever fully understand political ideology or the forces which motivate it, absent a grasp of r/K Selection Theory.

References

- Odum, Eugene P. (1959). Fundamentals of Ecology (Second ed.). Philadelphia and London: W. B. Saunders Co. p. 546 p. ISBN 9780721669410. OCLC 554879

- Tobiason, D. M.; Seifert, H. S. (19 February 2010). «Genomic Content of Neisseria Species». Journal of Bacteriology. 192 (8): 2160–2168. doi:10.1128/JB.01593-09. PMC 2849444

- Begon, M.; Townsend, C. R.; Harper, J. L. (2006). Ecology: From Individuals to Ecosystems (4th ed.). Oxford, UK: Blackwell Publishing. ISBN 978-1-4051-1117-1

- Hassell, Michael P. (June 1980). "Foraging Strategies, Population Models and Biological Control: A Case Study". The Journal of Animal Ecology. 49 (2): 603. doi:10.2307/4267

- Whitham, T. G. (1978). "Habitat Selection by Pemphigus Aphids in Response to Response Limitation and Competition". Ecology. Ecology, Vol. 59, No. 6. 59 (6): 1164–1176. doi:10.2307/1938230. JSTOR 1938230

- Chandler, M.; Bird, R.E.; Caro, L. (May 1975). "The replication time of the Escherichia coli K12 chromosome as a function of cell doubling time". Journal of Molecular Biology. 94 (1): 127–132. doi:10.1016/0022-2836(75)90410-6

- McKeown, Thomas (1976). The Modern Rise of Population. London, UK: Edward Arnold. ISBN 9780713159868

- Levins, R. (1970). Gerstenhaber, M., ed. Extinction. In: Some Mathematical Questions in Biology. AMS Bookstore. pp. 77–107. ISBN 978-0-8218-1152-8

- Weiss, Volkmar (2007). "The Population Cycle Drives Human History - from a Eugenic Phase into a Dysgenic Phase and Eventual Collapse". The Journal of Social, Political and Economic Studies. 32 (3): 327–58

- Hanski, I. (1998). "Metapopulation dynamics" (PDF). Nature. 396 (6706): 41–49. doi:10.1038/23876. Archived from the original (PDF) on 2010-12-31

- Gillespie, JH (2001). «Is the population size of a species relevant to its evolution?». Evolution. 55 (11): 2161–2169. doi:10.1111/j.0014-3820.2001.tb00732.x. PMID 11794777

- Introduction to Social Macrodynamics: Compact Macromodels of the World System Growth by Andrey Korotayev, Artemy Malkov, and Daria Khaltourina. ISBN 5-484-00414-4

Chapter 4

Understanding the Diverse Biogeochemical Cycles

A biogeochemical cycle refers to the cyclic pathway by which chemical substances move through the biotic and abiotic compartments of the Earth. Carbon, oxygen, sulfur, phosphorus, nitrogen and water cycles are significant processes of the Earth, which have been discussed in detail in this chapter.

Biogeochemical Cycle

A biogeochemical cycle is one of several natural cycles, in which conserved matter moves through the biotic and abiotic parts of an ecosystem.

In biology, conserved matter refers to the finite amount of matter, in the form of atoms, that is present within the Earth. Since, according to the Law of Conservation of Mass, matter cannot be created or destroyed, all atoms of matter are cycled through Earth's systems albeit in various forms.

In other words, the Earth only receives energy from the sun, which is given off as heat, whilst all other chemical elements remain within a closed system.

The main chemical elements that are cycled are: carbon (C), hydrogen (H), nitrogen (N), oxygen (O), phosphorous (P) and sulfur (S). These are the building blocks of life, and are used for essential processes, such as metabolism, the formation of amino acids, cell respiration and the building of tissues.

These fundamental elements can be easily remembered with the acronym CHNOPS.

Each of these elements is circulated through the biotic components, which are the living parts of an ecosystem, and the abiotic components, which are the non-living parts.

The abiotic components can be subdivided into three categories: thehydrosphere (water), the atmosphere (air) and the lithosphere(rock).

The biosphere is a term which can be used to describe the system that contains all living organisms, including plants, animals and bacteria, as well as their interactions among and between each other, and their interactions with the Earth's abiotic systems. The biosphere is sometimes called the ecosphere, and can be defined as the sum of all ecosystems.

With this knowledge, the words "biogeochemical cycle" can be easily broken down. "Bio-" is the biotic system, "geo-" is the geological component, and "chemical" is the elements which are moved through a "cycle".

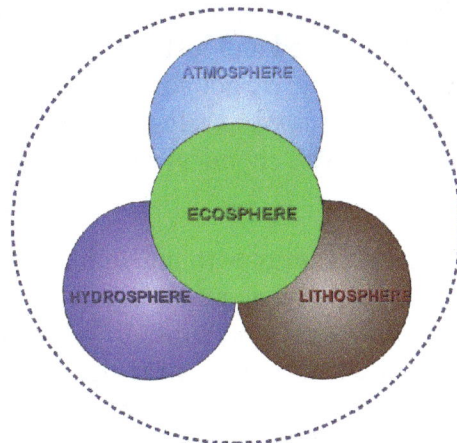

Biosphere system

At particular stages of their cycling, any of the elements may be stored and accumulated within a particular place for a long period time (e.g. within a rocky substrate, or in the atmosphere). These places are called "sinks" or "reservoirs".

A "source" is anything from which an element is output, for example volcanoes give off large amounts of carbon in the form of CO_2, while human waste is a source for nitrogen, sulfur and phosphorous.

Systems

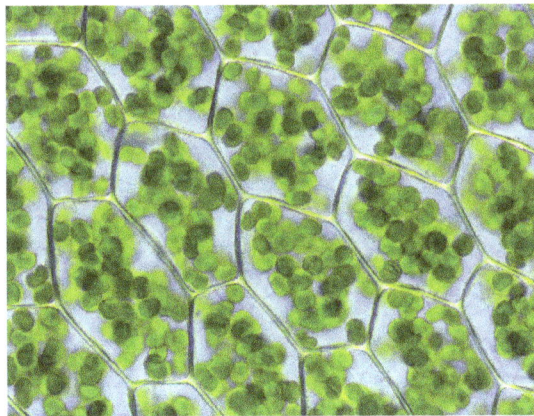

Chloroplasts conduct photosynthesis and are found in plant cells and other eukaryoticorganisms.
These are Chloroplasts visible in the cells of Plagiomnium affine — Many-fruited Thyme-moss

Ecological systems (ecosystems) have many biogeochemical cycles operating as a part of the system, for example the water cycle, the carbon cycle, the nitrogen cycle, etc. All chemical elements occurring in organisms are part of biogeochemical cycles. In addition to being a part of living organisms, these chemical elements also cycle through abiotic factors of ecosystems such as water (hydrosphere), land (lithosphere), and/or the air (atmosphere).

The living factors of the planet can be referred to collectively as the biosphere. All the nutrients—such as carbon, nitrogen, oxygen, phosphorus, and sulfur—used in ecosystems by living organisms are a part of a closed system; therefore, these chemicals are recycled instead of being lost and replenished constantly such as in an open system.

The flow of energy in an ecosystem is an open system; the sun constantly gives the planet energy in the form of light while it is eventually used and lost in the form of heat throughout the trophic levels of a food web. Carbon is used to make carbohydrates, fats, and proteins, the major sources of food energy. These compounds are oxidized to release carbon dioxide, which can be captured by plants to make organic compounds. The chemical reaction is powered by the light energy of the sun.

It is possible for an ecosystem to obtain energy without sunlight. Carbon must be combined with hydrogen and oxygen in order to be utilized as an energy source, and this process depends on sunlight. Ecosystems in the deep sea, where no sunlight can penetrate, use sulfur. Hydrogen sulfide near hydrothermal vents can be utilized by organisms such as the giant tube worm. In the sulfur cycle, sulfur can be forever recycled as a source of energy. Energy can be released through the oxidation and reduction of sulfur compounds (e.g., oxidizing elemental sulfur to sulfite and then to sulfate).

Although the Earth constantly receives energy from the sun, its chemical composition is essentially fixed, as additional matter is only occasionally added by meteorites. Because this chemical composition is not replenished like energy, all processes that depend on these chemicals must be recycled. These cycles include both the living biosphere and the nonliving lithosphere, atmosphere, and hydrosphere.

Reservoirs

Coal is a reservoir of carbon

The chemicals are sometimes held for long periods of time in one place. This place is called a reservoir, which, for example, includes such things as coaldeposits that are storing carbon for a long period of time. When chemicals are held for only short periods of time, they are being held in exchange pools. Examples of exchange pools include plants and animals.

Plants and animals utilize carbon to produce carbohydrates, fats, and proteins, which can then be used to build their internal structures or to obtain energy. Plants and animals temporarily use carbon in their systems and then release it back into the air or surrounding medium. Generally, reservoirs are abiotic factors whereas exchange pools are biotic factors. Carbon is held for a relatively short time in plants and animals in comparison to coal deposits. The amount of time that a chemical is held in one place is called its residence.

Important Cycles

The most well-known and important biogeochemical cycles are shown below:

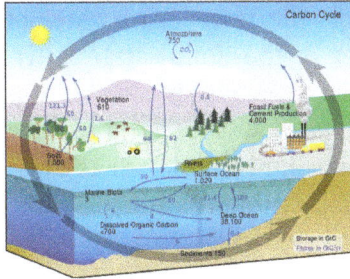

Carbon cycle

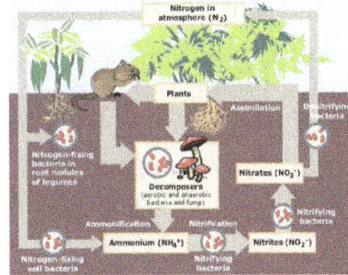

Nitrogen cycle

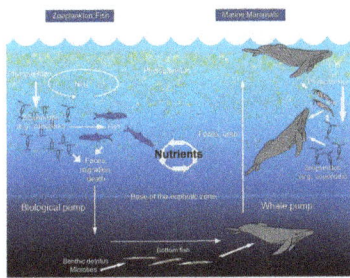

Nutrient cycle

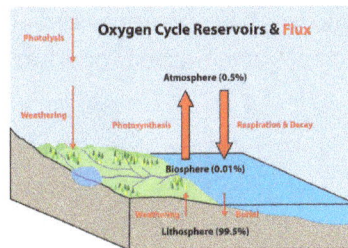

Oxygen cycle

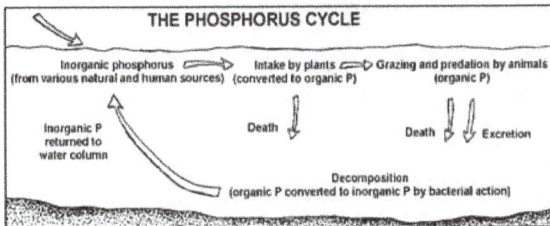

Phosphorus cycle

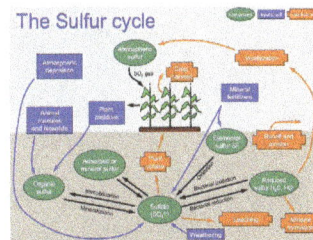

Sulfur cycle

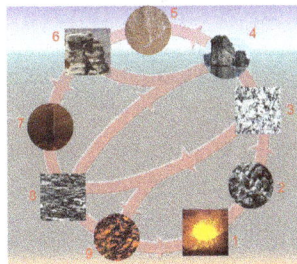

Rock cycle

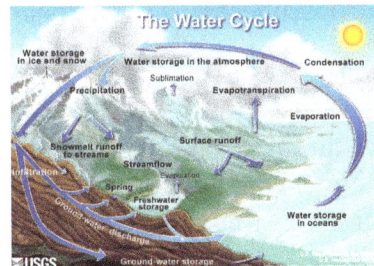

Water cycle

Water Cycle

Water cycle, also called hydrologic cycle, cycle that involves the continuous circulation of water in the Earth-atmosphere system. Of the many processes involved in the water cycle, the most important are evaporation, transpiration, condensation, precipitation, and runoff. Although the total amount of water within the cycle remains essentially constant, its distribution among the various processes is continually changing.

Evaporation, one of the major processes in the cycle, is the transfer of water from the surface of the Earth to the atmosphere. By evaporation, water in the liquid state is transferred to the gaseous, or vapour, state. This transfer occurs when some molecules in a water mass have attained sufficient kinetic energy to eject themselves from the water surface. The main factors affecting evaporation are temperature, humidity, wind speed, and solar radiation. The direct measurement of evaporation, though desirable, is difficult and possible only at point locations. The principal source of water vapour is the oceans, but evaporation also occurs in soils, snow, and ice. Evaporation from snow and ice, the direct conversion from solid to vapour, is known as sublimation. Transpiration is the evaporation of water through minute pores, or stomata, in the leaves of plants. For practical purposes, transpiration and the evaporation from all water, soils, snow, ice, vegetation, and other surfaces are lumped together and called evapotranspiration, or total evaporation.

The water cycle begins and ends in the ocean.Created and produced by QA International

Water vapour is the primary form of atmospheric moisture. Although its storage in the atmosphere is comparatively small, water vapour is extremely important in forming the moisture supply for dew, frost, fog, clouds, and precipitation. Practically all water vapour in the atmosphere is confined to the troposphere (the region below 6 to 8 miles [10 to 13 km] altitude).

The transition process from the vapour state to the liquid state is called condensation. Condensation may take place as soon as the air contains more water vapour than it can receive from a free water surface through evaporation at the prevailing temperature. This condition occurs as the consequence of either cooling or the mixing of air masses of different temperatures. By condensation, water vapour in the atmosphere is released to form precipitation.

Precipitation that falls to the Earth is distributed in four main ways: some is returned to the atmosphere by evaporation, some may be intercepted by vegetation and then evaporated from the surface of leaves, some percolates into the soil by infiltration, and the remainder flows directly as surface runoff into the sea. Some of the infiltrated precipitation may later percolate into streams as groundwater runoff. Direct measurement of runoff is made by stream gauges and plotted against time on hydrographs.

Most groundwater is derived from precipitation that has percolated through the soil. Groundwater flow rates, compared with those of surface water, are very slow and variable, ranging from a few millimetres to a few metres a day. Groundwater movement is studied by tracer techniques and remote sensing.

Ice also plays a role in the water cycle. Ice and snow on the Earth's surface occur in various forms such as frost, sea ice, and glacier ice. When soil moisture freezes, ice also occurs beneath the Earth's surface, forming permafrost in tundra climates. About 18,000 years ago glaciers and ice caps covered approximately one-third of the Earth's land surface. Today about 12 percent of the land surface remains covered by ice masses.

Description

The sun, which drives the water cycle, heats water in oceans and seas. Water evaporates as water vapor into the air. Some ice and snow sublimates directly into water vapor. Evapotranspiration is water transpired from plants and evaporated from the soil. The water molecule H_2O has smaller molecular mass than the major components of the atmosphere, nitrogen and oxygen, N_2 and O_2, hence is less dense. Due to the significant difference in density, buoyancy drives humid air higher. As altitude increases, air pressure decreases and the temperature drops. The lower temperature causes water vapor to condense into tiny liquid water droplets which are heavier than the air, and fall unless supported by an updraft. A huge concentration of these droplets over a large space up in the atmosphere become visible as cloud. Some condensation is near ground level, and called fog.

Atmospheric circulation moves water vapor around the globe, cloud particles collide, grow, and fall out of the upper atmospheric layers as precipitation. Some precipitation falls as snow or hail, sleet, and can accumulate as ice caps and glaciers, which can store frozen water for thousands of years. Most water falls back into the oceans or onto land as rain, where the water flows over the ground as surface runoff. A portion of runoff enters rivers in valleys in the landscape, with streamflow moving water towards the oceans. Runoff and water emerging from the ground (groundwater) may be stored as freshwater in lakes. Not all runoff flows into rivers, much of it soaks into the ground as infiltration. Some water infiltrates deep into the ground and replenishes aquifers, which can store freshwater for long periods of time. Some infiltration stays close to the land surface and can seep back into surface-water bodies (and the ocean) as groundwater discharge. Some groundwater finds openings in the land surface and comes out as freshwater springs. In river valleys and floodplains, there is often continuous water exchange between surface water and ground water in the hyporheic zone. Over time, the water returns to the ocean, to continue the water cycle.

Processes

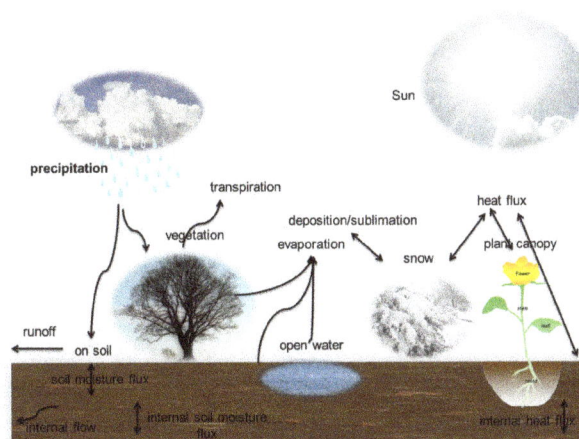

Many different processes lead to movements and phase changes in water

Precipitation

Condensed water vapor that falls to the Earth's surface. Most precipitation occurs as rain, but also includes snow, hail, fog drip, graupel, and sleet. Approximately 505,000 km³ (121,000 cu mi) of water falls as precipitation each year, 398,000 km³ (95,000 cu mi) of it over the oceans. The rain on land contains 107,000 km³ (26,000 cu mi) of water per year and a snowing only 1,000 km³ (240 cu mi). 78% of global precipitation occurs over the ocean.

Canopy interception

The precipitation that is intercepted by plant foliage eventually evaporates back to the atmosphere rather than falling to the ground.

Snowmelt

The runoff produced by melting snow.

Runoff

The variety of ways by which water moves across the land. This includes both surface runoff and channel runoff. As it flows, the water may seep into the ground, evaporate into the air, become stored in lakes or reservoirs, or be extracted for agricultural or other human uses.

Infiltration

The flow of water from the ground surface into the ground. Once infiltrated, the water becomes soil moisture or groundwater. A recent global study using water stable isotopes, however, shows that not all soil moisture is equally available for groundwater recharge or for plant transpiration.

Subsurface flow

The flow of water underground, in the vadose zone and aquifers. Subsurface water may return to the surface (e.g. as a spring or by being pumped) or eventually seep into the oceans. Water returns to the land surface at lower elevation than where it infiltrated, under the force of gravity or gravity induced pressures. Groundwater tends to move slowly and is replenished slowly, so it can remain in aquifers for thousands of years.

Evaporation

The transformation of water from liquid to gas phases as it moves from the ground or bodies of water into the overlying atmosphere. The source of energy for evaporation is primarily solar radiation. Evaporation often implicitly includes transpiration from plants, though together they are specifically referred to as evapotranspiration. Total annual evapotranspiration amounts to approximately 505,000 km³ (121,000 cu mi) of water, 434,000 km³ (104,000 cu mi) of which evaporates from the oceans. 86% of global evaporation occurs over the ocean.

Sublimation

The state change directly from solid water (snow or ice) to water vapor by passing the liquid state.

Deposition

This refers to changing of water vapor directly to ice.

Advection

The movement of water through the atmosphere. Without advection, water that evaporated over the oceans could not precipitate over land.

Condensation

The transformation of water vapor to liquid water droplets in the air, creating clouds and fog.

Transpiration

The release of water vapor from plants and soil into the air.

Percolation

Water flows vertically through the soil and rocks under the influence of gravity.

Plate tectonics

Water enters the mantle via subduction of oceanic crust. Water returns to the surface via volcanism.

The water cycle involves many of these processes.

Residence Times

The residence time of a reservoir within the hydrologic cycle is the average time a water molecule will spend in that reservoir. It is a measure of the average age of the water in that reservoir.

Average reservoir residence times	
Reservoir	Average residence time
Antarctica	20,000 years
Oceans	3,200 years
Glaciers	20 to 100 years
Seasonal snow cover	2 to 6 months
Soil moisture	1 to 2 months
Groundwater: shallow	100 to 200 years
Groundwater: deep	10,000 years
Lakes (see lake retention time)	50 to 100 years
Rivers	2 to 6 months
Atmosphere	9 days

Groundwater can spend over 10,000 years beneath Earth's surface before leaving. Particularly old groundwater is called fossil water. Water stored in the soil remains there very briefly, because it is spread thinly across the Earth, and is readily lost by evaporation, transpiration, stream flow, or groundwater recharge. After evaporating, the residence time in the atmosphere is about 9 days before condensing and falling to the Earth as precipitation.

The major ice sheets – Antarctica and Greenland – store ice for very long periods. Ice from Antarctica has been reliably dated to 800,000 years before present, though the average residence time is shorter.

In hydrology, residence times can be estimated in two ways. The more common method relies on the principle of conservation of mass and assumes the amount of water in a given reservoir is roughly constant. With this method, residence times are estimated by dividing the volume of the reservoir by the rate by which water either enters or exits the reservoir. Conceptually, this is equivalent to timing how long it would take the reservoir to become filled from empty if no water were to leave (or how long it would take the reservoir to empty from full if no water were to enter).

An alternative method to estimate residence times, which is gaining in popularity for dating groundwater, is the use of isotopic techniques. This is done in the subfield of isotope hydrology.

Changes Over Time

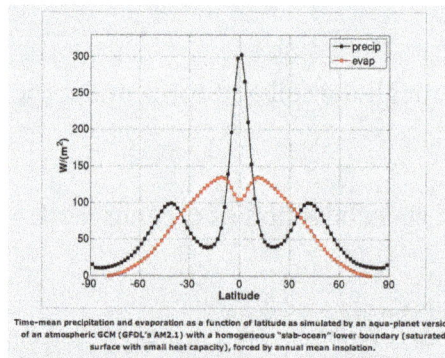

Time-mean precipitation and evaporation as a function of latitude as simulated by an aqua-planet version of an atmospheric GCM (GFDL's AM2.1) with a homogeneous "slab-ocean" lower boundary (saturated surface with small heat capacity), forced by annual mean insolation

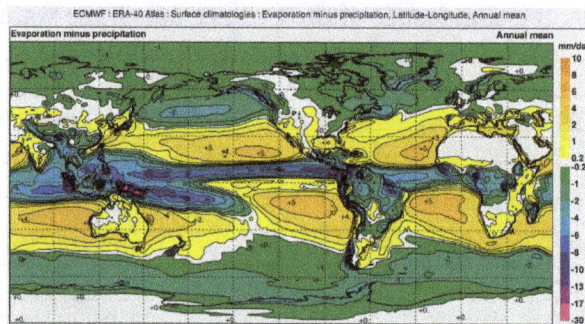

Global map of annual mean evaporation minus precipitation by latitude-longitude

The water cycle describes the processes that drive the movement of water throughout the hydrosphere. However, much more water is "in storage" for long periods of time than is actually moving through the cycle. The storehouses for the vast majority of all water on Earth are the oceans. It is estimated that of the 332,500,000 mi³ (1,386,000,000 km³) of the world's water supply, about 32 1,000,000 mi³ (1,338,000,000 km³) is stored in oceans, or about 97%. It is also estimated that the oceans supply about 90% of the evaporated water that goes into the water cycle.

During colder climatic periods, more ice caps and glaciers form, and enough of the global water supply accumulates as ice to lessen the amounts in other parts of the water cycle. The reverse is true during warm periods. During the last ice age, glaciers covered almost one-third of Earth's land mass with the result being that the oceans were about 122 m (400 ft) lower than today. During the last global "warm spell," about 125,000 years ago, the seas were about 5.5 m (18 ft) higher than they are now. About three million years ago the oceans could have been up to 50 m (165 ft) higher.

The scientific consensus expressed in the 2007 Intergovernmental Panel on Climate Change (IPCC) Summary for Policymakers is for the water cycle to continue to intensify throughout the 21st century, though this does not mean that precipitation will increase in all regions. In subtropical land areas — places that are already relatively dry — precipitation is projected to decrease during the 21st century, increasing the probability of drought. The drying is projected to be strongest near the poleward margins of the subtropics (for example, the Mediterranean Basin, South Africa, southern Australia, and the Southwestern United States). Annual precipitation amounts are expected to increase in near-equatorial regions that tend to be wet in the present climate, and also at high latitudes. These large-scale patterns are present in nearly all of the climate model simulations conducted at several international research centers as part of the 4th Assessment of the IPCC. There is now ample evidence that increased hydrologic variability and change in climate has and will continue to have a profound impact on the water sector through the hydrologic cycle, water availability, water demand, and water allocation at the global, regional, basin, and local levels. Research published in 2012 in Sciencebased on surface ocean salinity over the period 1950 to 2000 confirm this projection of an intensified global water cycle with salty areas becoming more saline and fresher areas becoming more fresh over the period:

> Fundamental thermodynamics and climate models suggest that dry regions will become drier and wet regions will become wetter in response to warming. Efforts to detect this long-term response in sparse surface observations of rainfall and evaporation remain ambiguous. We show that ocean salinity patterns express an identifiable fingerprint of an intensifying water cycle. Our 50-year observed global surface salinity changes, combined with changes from global climate models, present robust evidence of an intensified global water cycle at a rate of 8 ± 5% per degree of surface warming. This rate is double the response projected by current-generation climate models and suggests that a substantial (16 to 24%) intensification of the global water cycle will occur in a future 2° to 3° warmer world.

An instrument carried by the SAC-D satellite Aquarius, launched in June, 2011, measured global sea surface salinity.

Glacial retreat is also an example of a changing water cycle, where the supply of water to glaciers from precipitation cannot keep up with the loss of water from melting and sublimation. Glacial retreat since 1850 has been extensive.

Human activities that alter the water cycle include:

- agriculture
- industry
- alteration of the chemical composition of the atmosphere
- construction of dams

- deforestation and afforestation

- removal of groundwater from wells

- water abstraction from rivers

- urbanization

Effects on Climate

The water cycle is powered from solar energy. 86% of the global evaporation occurs from the oceans, reducing their temperature by evaporative cooling. Without the cooling, the effect of evaporation on the greenhouse effect would lead to a much higher surface temperature of 67 °C (153 °F), and a warmer planet.

Aquifer drawdown or overdrafting and the pumping of fossil water increases the total amount of water in the hydrosphere, and has been postulated to be a contributor to sea-level rise.

Effects on Biogeochemical Cycling

While the water cycle is itself a biogeochemical cycle, flow of water over and beneath the Earth is a key component of the cycling of other biogeochemicals.Runoff is responsible for almost all of the transport of eroded sediment and phosphorus from land to waterbodies. The salinity of the oceans is derived from erosion and transport of dissolved salts from the land. Cultural eutrophication of lakes is primarily due to phosphorus, applied in excess to agricultural fields in fertilizers, and then transported overland and down rivers. Both runoff and groundwater flow play significant roles in transporting nitrogen from the land to waterbodies. The dead zone at the outlet of the Mississippi River is a consequence of nitrates from fertilizer being carried off agricultural fields and funnelled down the river system to the Gulf of Mexico. Runoff also plays a part in the carbon cycle, again through the transport of eroded rock and soil.

Slow Loss over Geologic Time

The hydrodynamic wind within the upper portion of a planet's atmosphere allows light chemical elements such as Hydrogen to move up to the exobase, the lower limit of the exosphere, where the gases can then reach escape velocity, entering outer space without impacting other particles of gas. This type of gas loss from a planet into space is known as planetary wind. Planets with hot lower atmospheres could result in humid upper atmospheres that accelerate the loss of hydrogen.

The water cycle has a dramatic influence on Earth's climate and ecosystems.

Climate is all the weather conditions of an area, evaluated over a period of time. Two weather conditions that contribute to climate include humidity and temperature. These weather conditions are influenced by the water cycle.

Humidity is simply the amount of water vapor in the air. As water vapor is not evenly distributed by the water cycle, some regions experience higher humidity than others. This contributes to radically different climates. Islands or coastal regions, where water vapor makes up more of the atmosphere, are usually much more humid than inland regions, where water vapor is scarcer.

A region's temperature also relies on the water cycle. Through the water cycle, heat is exchanged and temperatures fluctuate. As water evaporates, for example, it absorbs energy and cools the local environment. As water condenses, it releases energy and warms the local environment.

The Water Cycle and the Landscape

The water cycle also influences the physical geography of the Earth. Glacial melt and erosion caused by water are two of the ways the water cycle helps create Earth's physical features.

As glaciers slowly expand across a landscape, they can carve away entire valleys, create mountain peaks, and leave behind rubble as big as boulders. Yosemite Valley, part of Yosemite National Park in the U.S. state of California, is a glacial valley. The famous Matterhorn, a peak on the Alps between Switzerland and Italy, was carved as glaciers collided and squeezed up the earth between them. Canada's "Big Rock" is one of the world's largest "glacial erratics," boulders left behind as a glacier advances or retreats.

Glacial melt can also create landforms. The Great Lakes, for example, are part of the landscape of the Midwest of the United States and Canada. The Great Lakes were created as an enormous ice sheet melted and retreated, leaving liquid pools.

The process of erosion and the movement of runoff also create varied landscapes across the Earth's surface. Erosion is the process by which earth is worn away by liquid water, wind, or ice.

Erosion can include the movement of runoff. The flow of water can help carve enormous canyons, for example. These canyons can be carved by rivers on high plateaus (such as the Grand Canyon, on the Colorado Plateau in the U.S. state of Arizona). They can also be carved by currents deep in the ocean (such as the Monterey Canyon, in the Pacific Ocean off the coast of the U.S. state of California).

Carbon Cycle

Carbon is a very important element, as it makes up organic matter, which is a part of all life. Carbon follows a certain route on earth, called the carbon cycle. Through following the carbon cycle we can also study energy flows on earth, because most of the chemical energy needed for life is stored in organic compounds as bonds between carbon atoms and other atoms. The carbon cycle naturally consists of two parts, the terrestrial and the aquatic carbon cycle. The aquatic carbon cycle is concerned with the movements of carbon through marine ecosystems and the terrestrial carbon cycle is concerned with the movement of carbon through terrestrial ecosystems.

The carbon cycle is based on carbon dioxide (CO_2), which can be found in air in the gaseous form, and in water in dissolved form. Terrestrial plants use atmospheric carbon dioxide from the atmosphere, to generate oxygen that sustains animal life. Aquatic plants also generate oxygen, but they use carbon dioxide from water. The process of oxygen generation is called photosynthesis. During photosynthesis, plants and other producers transfer carbon dioxide and water into complex carbohydrates, such as glucose, under the influence of sunlight. Only plants and some bacteria have the ability to conduct this process, because they possess chlorophyll; a pigment molecule in leaves that they can capture solar energy with.

The overall reaction of photosynthesis is: carbon dioxide + water + solar energy -> glucose + oxygen 6 CO_2 + 6 H_2O + solar energy -> $C_6H_{12}O_6$ + 6 O_2

The oxygen that is produced during photosynthesis will sustain non-producing life forms, such as animals, and most micro organisms. Animals are called consumers, because they use the oxygen that is produced by plants. Carbon dioxide is released back into the atmosphere during respiration of consumers, which breaks down glucose and other complex organic compounds and converts the carbon back to carbon dioxide for reuse by producers.

Carbon that is used by producers, consumers and decomposers cycles fairly rapidly through air, water and biota. But carbon can also be stored as biomass in the roots of trees and other organic matter for many decades. This carbon is released back into the atmosphere by decomposition, as was noted before. Not all organic matter is immediately decomposed. Under certain conditions dead plant matter accumulates faster than it is decomposed within an ecosystem. The remains are locked away in underground deposits. When layers of sediment compress this matter fossil fuels will be formed, after many centuries. Long-term geological processes may expose the carbon in these fuels to air after a long period of time, but usually the carbon within the fossil fuels is released during humane combustion processes. The combustion of fossil fuels has supplied us with energy for as long as we can remember. But the human population of the world has been expanding and so has our demand for energy. That is why fossil fuels are burned very extensively. This is not without consequences, because we are burning fossil fuels much faster than they develop. Because of our actions fossil fuels have become non-renewable recourses.

Although the combustion of fossil fuels mainly adds carbon dioxide to air, some of it is also released during natural processes, such as volcanic eruptions.

In the aquatic ecosystem carbon dioxide can be stored in rocks and sediments. It will take a long time before this carbon dioxide will be released, through weathering of rocks or geologic processes that bring sediment to the surface of water. Carbon dioxide that is stored in water will be present as either carbonate or bicarbonate ions. These ions are an important part of natural buffers that prevent the water from becoming too acidic or too basic. When the sun warms up the water carbonate and bicarbonate ions will be returned to the atmosphere as carbon dioxide.

Main Components

Carbon pools in the major reservoirs on earth	
Pool	Quantity (gigatons)
Atmosphere	720
Ocean (total)	38,400
Total inorganic	37,400
Total organic	1,000
Surface layer	670
Deep layer	36,730
Lithosphere	
Sedimentary carbonates	> 60,000,000
Kerogens	15,000,000
Terrestrial biosphere (total)	2,000
Living biomass	600 - 1,000
Dead biomass	1,200

Aquatic biosphere	1 - 2
Fossil fuels (total)	4,130
Coal	3,510
Oil	230
Gas	140
Other (peat)	250

The global carbon cycle is now usually divided into the following major reservoirs of carbon inter-connected by pathways of exchange:

- The atmosphere

- The terrestrial biosphere

- The ocean, including dissolved inorganic carbon and living and non-living marine biota

- The sediments, including fossil fuels, fresh water systems and non-living organic material.

- The Earth's interior (mantle and crust. These carbon stores interact with the other components through geological processes.

The carbon exchanges between reservoirs occur as the result of various chemical, physical, geological, and biological processes. The ocean contains the largest active pool of carbon near the surface of the Earth. The natural flows of carbon between the atmosphere, ocean, terrestrial ecosystems, and sediments is fairly balanced, so that carbon levels would be roughly stable without human influence.

Atmosphere

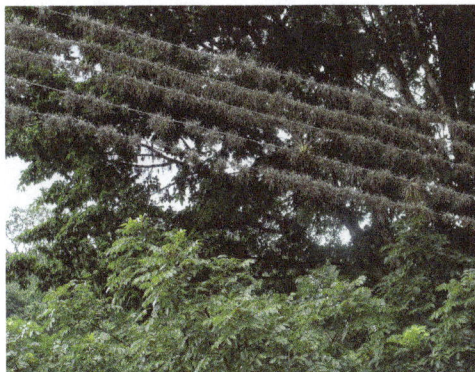

Epiphytes on electric wires. This kind of plant takes both CO_2 and water from the atmosphere for living and growing.

Carbon in the Earth's atmosphere exists in two main forms: carbon dioxide and methane. Both of these gases absorb and retain heat in the atmosphere and are partially responsible for the greenhouse effect. Methane produces a larger greenhouse effect per volume as compared to carbon dioxide, but it exists in much lower concentrations and is more short-lived than carbon dioxide, making carbon dioxide the more important greenhouse gas of the two.

Carbon dioxide is removed from the atmosphere primarily through photosynthesis and enters the terrestrial and oceanic biospheres. Carbon dioxide also dissolves directly from the atmosphere into bodies of water (ocean, lakes, etc.), as well as dissolving in precipitation as raindrops fall through the atmosphere. When dissolved in water, carbon dioxide reacts with water molecules and forms carbonic acid, which contributes to ocean acidity. It can then be absorbed by rocks through weathering. It also can acidify other surfaces it touches or be washed into the ocean.

Human activities over the past two centuries have significantly increased the amount of carbon in the atmosphere, mainly in the form of carbon dioxide, both by modifying ecosystems' ability to extract carbon dioxide from the atmosphere and by emitting it directly, e.g., by burning fossil fuels and manufacturing concrete.

Terrestrial Biosphere

The terrestrial biosphere includes the organic carbon in all land-living organisms, both alive and dead, as well as carbon stored in soils. About 500 gigatons of carbon are stored above ground in plants and other living organisms, while soil holds approximately 1,500 gigatons of carbon. Most carbon in the terrestrial biosphere is organic carbon, while about a third of soil carbon is stored in inorganic forms, such as calcium carbonate. Organic carbon is a major component of all organisms living on earth. Autotrophs extract it from the air in the form of carbon dioxide, converting it into organic carbon, while heterotrophs receive carbon by consuming other organisms.

A portable soil respiration system measuring soil CO_2 flux

Because carbon uptake in the terrestrial biosphere is dependent on biotic factors, it follows a diurnal and seasonal cycle. In CO_2 measurements, this feature is apparent in the Keeling curve. It is strongest in the northern hemisphere, because this hemisphere has more land mass than the southern hemisphere and thus more room for ecosystems to absorb and emit carbon.

Carbon leaves the terrestrial biosphere in several ways and on different time scales. The combustion or respiration of organic carbon releases it rapidly into the atmosphere. It can also be exported into the ocean through rivers or remain sequestered in soils in the form of inert carbon. Carbon stored in soil can remain there for up to thousands of years before being washed into rivers by erosion or released into the atmosphere through soil respiration. Between 1989 and 2008 soil respiration increased by about 0.1% per year. In 2008, the global total of CO_2 released by soil respiration was roughly 98 billion tonnes, about 10 times more carbon than humans are now putting into the atmosphere each year by burning fossil fuel (this does not represent a net transfer of carbon from soil to atmosphere, as the respiration is largely offset by inputs to soil carbon). There are a few plausible explanations for this trend, but the most likely explanation is that increasing temperatures have increased rates of decomposition of soil organic matter, which has increased the flow of CO_2. The length of carbon sequestering in soil is dependent on local climatic conditions and thus changes in the course of climate change.

Ocean

The ocean can be conceptually divided into a surface layer within which water makes frequent (daily to annual) contact with the atmosphere, and a deep layer below the typical mixed layer depth of a few hundred meters or less, within which the time between consecutive contacts may be centuries. The dissolved inorganic carbon (DIC) in the surface layer is exchanged rapidly with the atmosphere, maintaining equilibrium. Partly because its concentration of DIC is about 15% higher but mainly due to its larger volume, the deep ocean contains far more carbon--it›s the largest pool of actively cycled carbon in the world, containing 50 times more than the atmosphere--but the timescale to reach equilibrium with the atmosphere is hundreds of years: the exchange of carbon between the two layers, driven by thermohaline circulation, is slow.

Carbon enters the ocean mainly through the dissolution of atmospheric carbon dioxide, a small fraction of which is converted into carbonate. It can also enter the ocean through rivers as dissolved organic carbon. It is converted by organisms into organic carbon through photosynthesis and can either be exchanged throughout the food chain or precipitated into the ocean›s deeper, more carbon rich layers as dead soft tissue or in shells as calcium carbonate. It circulates in this layer for long periods of time before either being deposited as sediment or, eventually, returned to the surface waters through thermohaline circulation.

Oceanic absorption of CO_2 is one of the most important forms of carbon sequestering limiting the human-caused rise of carbon dioxide in the atmosphere. However, this process is limited by a number of factors. CO_2 absorption makes water more acidic, which affects ocean biosystems. The projected rate of increasing oceanic acidity could slow the biological precipitation of calcium carbonates, thus decreasing the ocean's capacity to absorb carbon dioxide.

Earth's Interior

The geologic component of the carbon cycle operates slowly in comparison to the other parts of the global carbon cycle. It is one of the most important determinants of the amount of carbon in the atmosphere, and thus of global temperatures.

Most of the earth's carbon is stored inertly in the earth's lithosphere. Much of the carbon stored in the earth›s mantle was stored there when the earth formed. Some of it was deposited in the form of organic carbon from the biosphere. Of the carbon stored in the geosphere, about 80% is limestone and its derivatives, which form from the sedimentation of calcium carbonate stored in the shells of marine organisms. The remaining 20% is stored as kerogens formed through the sedimentation and burial of terrestrial organisms under high heat and pressure. Organic carbon stored in the geosphere can remain there for millions of years.

Carbon can leave the geosphere in several ways. Carbon dioxide is released during the metamorphosis of carbonate rocks when they are subducted into the earth›s mantle. This carbon dioxide can be released into the atmosphere and ocean through volcanoes and hotspots. It can also be removed by humans through the direct extraction of kerogens in the form of fossil fuels. After extraction, fossil fuels are burned to release energy, thus emitting the carbon they store into the atmosphere.

Human Influence

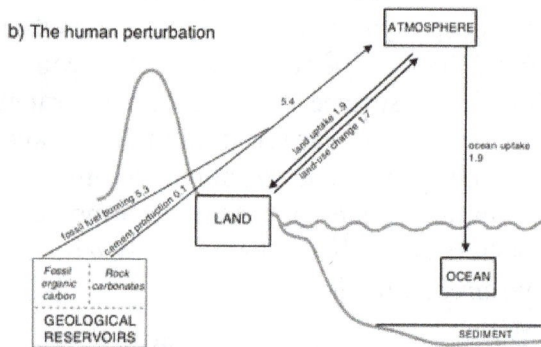

Human activity since the industrial era has changed the balance in the natural carbon cycle. Units are in gigatons.

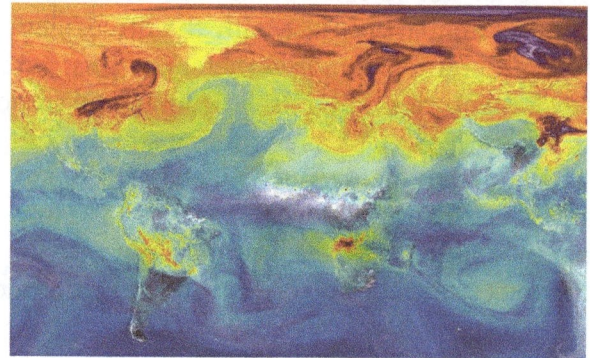

CO$_2$ in Earth's atmosphere if half of global-warming emissions are notabsorbed.

Since the industrial revolution, human activity has modified the carbon cycle by changing its components' functions and directly adding carbon to the atmosphere.

The largest human impact on the carbon cycle is through direct emissions from burning fossil fuels, which transfers carbon from the geosphere into the atmosphere. The rest of this increase is caused mostly by changes in land-use, particularly deforestation.

Another direct human impact on the carbon cycle is the chemical process of calcination of limestone for clinker production, which releases CO$_2$. Clinker is an industrial precursor of cement.

Humans also influence the carbon cycle indirectly by changing the terrestrial and oceanic biosphere. Over the past several centuries, direct and indirect human-caused land use and land cover change (LUCC) has led to the loss of biodiversity, which lowers ecosystems' resilience to environmental stresses and decreases their ability to remove carbon from the atmosphere. More directly, it often leads to the release of carbon from terrestrial ecosystems into the atmosphere. Deforestation for agricultural purposes removes forests, which hold large amounts of carbon, and replaces them, generally with agricultural or urban areas. Both of these replacement land cover types store comparatively small amounts of carbon, so that the net product of the process is that more carbon stays in the atmosphere.

Other human-caused changes to the environment change ecosystems' productivity and their ability to remove carbon from the atmosphere. Air pollution, for example, damages plants and soils, while many agricultural and land use practices lead to higher erosion rates, washing carbon out of soils and decreasing plant productivity.

Humans also affect the oceanic carbon cycle. Current trends in climate change lead to higher ocean temperatures, thus modifying ecosystems. Also, acid rain and polluted runoff from agriculture and industry change the ocean's chemical composition. Such changes can have dramatic effects on highly sensitive ecosystems such as coral reefs, thus limiting the ocean's ability to absorb carbon from the atmosphere on a regional scale and reducing oceanic biodiversity globally.

Arctic methane emissions indirectly caused by anthropogenic global warming also affect the carbon cycle, and contribute to further warming in what is known as climate change feedback.

On 12 November 2015, NASA scientists reported that human-made carbon dioxide (CO_2) continues to increase, reaching levels not seen in hundreds of thousands of years: currently, the rate carbon dioxide release by the burning of fossil fuels is about double the net uptake by vegetation and the ocean.

The carbon cycle is most easily studied as two interconnected subcycles:

- One dealing with rapid carbon exchange among living organisms

- One dealing with long-term cycling of carbon through geologic processes

Although we will look at them separately, it's important to realize these cycles are linked. For instance, the same pools of atmospheric and oceanic CO_2 that are utilized by organisms are also fed and depleted by geological processes.

As a brief overview, carbon exists in the air largely as carbon dioxide—CO_2—gas, which dissolves in water and reacts with water molecules to produce bicarbonate—HCO_3^-. Photosynthesis by land plants, bacteria, and algae converts carbon dioxide or bicarbonate into organic molecules. Organic molecules made by photosynthesizers are passed through food chains, and cellular respiration converts the organic carbon back into carbon dioxide gas.

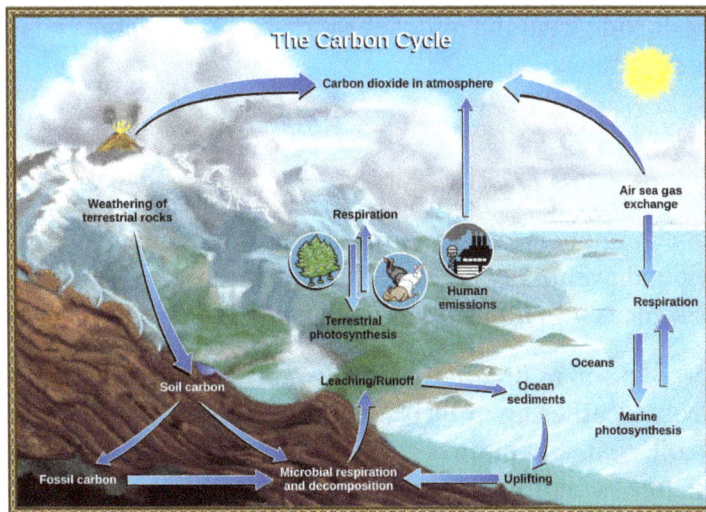

Longterm storage of organic carbon occurs when matter from living organisms is buried deep underground or sinks to the bottom of the ocean and forms sedimentary rock. Volcanic activity and, more recently, human burning of fossil fuels bring this stored carbon back into the carbon cycle. Although the formation of fossil fuels happens on a slow, geologic timescale, human release of the carbon they contain—as CO_2—is on a very fast timescale.

The Biological Carbon Cycle

Carbon enters all food webs, both terrestrial and aquatic, through autotrophs, or self-feeders. Almost all of these autotrophs are photosynthesizers, such as plants or algae.

Autotrophs capture carbon dioxide from the air or bicarbonate ions from the water and use them to make organic compounds such as glucose. Heterotrophs, or other-feeders, such as humans, consume the organic molecules, and the organic carbon is passed through food chains and webs.

How does carbon cycle back to the atmosphere or ocean? To release the energy stored in carbon-containing molecules, such as sugars, autotrophs and heterotrophs break these molecules down in a process called cellular respiration. In this process, the carbons of the molecule are released as carbon dioxide. Decomposers also release organic compounds and carbon dioxide when they break down dead organisms and waste products.

Carbon can cycle quickly through this biological pathway, especially in aquatic ecosystems. Overall, an estimated 1,000 to 100,000 million metric tons of carbon move through the biological pathway each year. For context, a metric ton is about the weight of an elephant or a small car!

The Geological Carbon Cycle

The geological pathway of the carbon cycle takes much longer than the biological pathway described above. In fact, it usually takes millions of years for carbon to cycle through the geological pathway. Carbon may be stored for long periods of time in the atmosphere, bodies of liquid water— mostly oceans— ocean sediment, soil, rocks, fossil fuels, and Earth's interior.

The level of carbon dioxide in the atmosphere is influenced by the reservoir of carbon in the oceans and vice versa. Carbon dioxide from the atmosphere dissolves in water and reacts with water molecules in the following reactions:

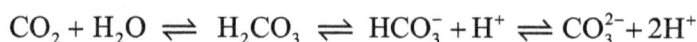

$$CO_2 + H_2O \rightleftharpoons H_2CO_3 \rightleftharpoons HCO_3^- + H^+ \rightleftharpoons CO_3^{2-} + 2H^+$$

The carbonate— CO_3^{2-} —released in this process combines with Ca^{2+} ions to make calcium carbonate— $CaCO_3$ —a key component of the shells of marine organisms. When the organisms die, their remains may sink and eventually become part of the sediment on the ocean floor. Over geologic time, the sediment turns into limestone, which is the largest carbon reservoir on Earth.

On land, carbon is stored in soil as organic carbon from the decomposition of living organisms or as inorganic carbon from weathering of terrestrial rock and minerals. Deeper under the ground are fossil fuels such as oil, coal, and natural gas, which are the remains of plants decomposed under anaerobic—oxygen-free—conditions. Fossil fuels take millions of years to form. When humans burn them, carbon is released into the atmosphere as carbon dioxide.

Another way for carbon to enter the atmosphere is by the eruption of volcanoes. Carbon-containing sediments in the ocean floor are taken deep within the Earth in a process called subduction, in which one tectonic plate moves under another. This process forms carbon dioxide, which can be released into the atmosphere by volcanic eruptions or hydrothermal vents.

Oxygen Cycle

The oxygen cycle elaborates how oxygen circulates in various forms through nature. Oxygen occurs freely in the air, trapped in the earth crust as chemical compounds, or dissolved in water. Oxygen in the atmosphere is about 21%, and it is the second most abundant gas after nitrogen. It is mostly utilized by living organisms, especially man and animals in respiration. Oxygen is also the most common element of human body.

Oxygen is also used during combustion, decomposition, and oxidation. The circulation of oxygen is through three main flow systems including the (air) atmosphere, the biosphere, and the earth's crust. In the oxygen cycle, the main driving factor is photosynthesis which is the process whereby green plants and algae make their own food by use of solar energy, water, and carbon dioxide to gives off oxygen as a by-product.

Hence, for oxygen to remain in the atmosphere, it has to circulate through various forms of nature which is essentially termed as the oxygen cycle. The circulation depends on the various activities on Earth.

Reservoirs

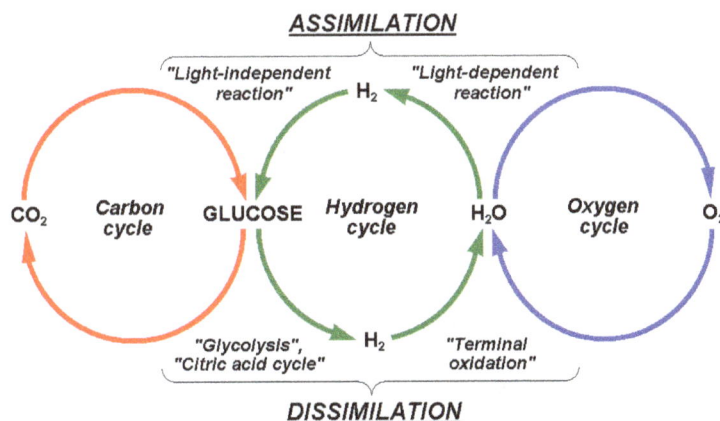

Interconnection between carbon, hydrogen and oxygen cycle in metabolism of photosynthesizing plants

By far the largest reservoir of Earth's oxygen is within the silicate and oxide minerals of the crust and mantle (99.5% by weight). The Earth›s atmosphere, hydrosphere and biosphere together weigh less than 0.05% of the Earth›s total mass. Oxygen is one of the most abundant elements on Earth and represents a large portion of each main reservoir:

- Atmosphere is 21% oxygen by volume present mainly as free oxygen molecules (O_2) with other oxygen-containing molecules including ozone (O_3), carbon dioxide (CO_2), water vapor (H_2O), and sulfur and nitrogen oxides (SO_2, NO, N_2O, etc.)

- Biosphere is 22% oxygen by volume present mainly as a component of organic molecules ($C_xH_xN_xO_x$) and water molecules

- Hydrosphere is 33% oxygen by volume present mainly as a component of water molecules with dissolved molecules including free oxygen and carbonic acids (H_xCO_3)

- Lithosphere is 94% oxygen by volume present mainly as silica minerals (SiO_2) and other oxide minerals

Movement of oxygen between the reservoirs is facilitated in large part by the presence of atmospheric free oxygen. The main source of atmospheric free oxygen is photosynthesis, which produces sugars and free oxygen from carbon dioxide and water:

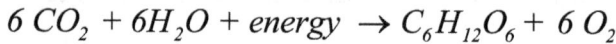

$$6\ CO_2 + 6H_2O + energy \rightarrow C_6H_{12}O_6 + 6\ O_2$$

Photosynthesizing organisms include the plant life of the land areas as well as the phytoplankton of the oceans. The tiny marine cyanobacterium Prochlorococcus was discovered in 1986 and accounts for more than half of the photosynthesis of the open ocean.

An additional source of atmospheric free oxygen comes from photolysis, whereby high-energy ultraviolet radiation breaks down atmospheric water and nitrous oxide into component atoms. The free H and N atoms escape into space, leaving O_2 in the atmosphere:

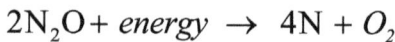

$$2H_2O + energy \rightarrow 4H + O_2$$

$$2N_2O + energy \rightarrow 4N + O_2$$

The main way free oxygen is lost from the atmosphere is via respiration and decay, mechanisms in which animal life and bacteria consume oxygen and release carbon dioxide.

The lithosphere also consumes atmospheric free oxygen by chemical weathering and surface reactions. An example of surface weathering chemistry is formation of iron oxides (rust):

$$4\ FeO + O_2 \rightarrow 2\ Fe_2O_3$$

Oxygen is also cycled between the biosphere and lithosphere. Marine organisms in the biosphere create calcium carbonate shell material ($CaCO_3$) that is rich in oxygen. When the organism dies, its shell is deposited on the shallow sea floor and buried over time to create the limestone sedimentary rock of the lithosphere. Weathering processes initiated by organisms can also free oxygen from the lithosphere. Plants and animals extract nutrient minerals from rocks and release oxygen in the process.

Capacities and Fluxes

The following tables offer estimates of oxygen cycle reservoir capacities and fluxes. These numbers are based primarily on estimates from (Walker, J. C. G.):

Table: Major reservoirs involved in the oxygen cycle

Reservoir	Capacity (kg O_2)	Flux in/out (kg O_2 per year)	Residence time (years)
Atmosphere	1.4×10^{18}	3×10^{14}	4500
Biosphere	1.6×10^{16}	3×10^{14}	50
Lithosphere	2.9×10^{20}	6×10^{11}	500000000

Table : Annual gain and loss of atmospheric oxygen (Units of 10^{10} kg O_2 per year)

Photosynthesis (land)	16,500
Photosynthesis (ocean)	13,500
Photolysis of N_2O	1.3
Photolysis of H_2O	0.03
Total gains	~ 30,000
Losses - respiration and decay	
Aerobic respiration	23,000
Microbial oxidation	5,100
Combustion of fossil fuel (anthropogenic)	1,200
Photochemical oxidation	600
Fixation of N_2 by lightning	12
Fixation of N_2 by industry (anthropogenic)	10
Oxidation of volcanic gases	5
Losses - weathering	
Chemical weathering	50
Surface reaction of O_3	12
Total losses	~ 30,000

Ozone

The presence of atmospheric oxygen has led to the formation of ozone (O_3) and the ozone layer within the stratosphere:

$$O_2 + uv\,light \rightarrow 2O \qquad (\lambda \lesssim 200\,nm)$$

$$O + O_2 \rightarrow O_3$$

The ozone layer is extremely important to modern life as it absorbs harmful ultraviolet radiation:

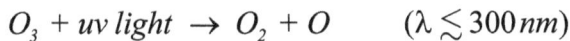

$$O_3 + uv\,light \rightarrow O_2 + O \qquad (\lambda \lesssim 300\,nm)$$

Process of Oxygen Cycle

The Atmosphere (air)

The atmosphere carries a small quantity of all oxygen, only about 0.35% of the entire earth's oxygen. In the atmosphere, oxygen is released by the process known as photolysis. Photolysis happens when the ultraviolet radiation of sunlight breaks apart oxygen-containing molecules such as nitrous oxide and atmospheric water to release free oxygen. The surplus oxygen recombines with other oxygen molecules to form ozone while the rest is freed into the atmosphere. Ozone is the layer that helps to shield the Earth from the dangerous ultra violet rays.

Biosphere

The biosphere carries the smallest quantity of all earth's oxygen, about 0.01%. In the biosphere, the major oxygen cycles are photosynthesis and respiration. In these two processes of the oxygen cycle, it is interconnected with the carbon cycle and the water cycle. During photosynthesis, plants and planktons use sunlight energy, water, and carbon dioxide to make food (carbohydrates) and release oxygen as a by-product. As such, plants and planktons are the main producers of oxygen in the ecosystem. They take in carbon dioxide and give out oxygen. Plants are estimated to replace about 99% of all the oxygen used.

On the other hand, respiration happens when humans and animals breathe in oxygen which is used during metabolism to break down carbohydrates and exhale carbon dioxide as a by-product. Such free carbon dioxide is then released into the environment and is used by plants and planktons during photosynthesis to give out molecular atmospheric oxygen, thus completing the oxygen cycle. Therefore, suffice is to say that oxygen enters organisms in the biosphere through respiration and is expelled through photosynthesis in a process that is interconnected with the carbon cycle-plus the water cycle.

However, the continued release of carbon dioxide into the atmosphere by burning fossil fuels and automobile pollution affects the oxygen cycle.

The Lithosphere (Earth's Crust)

The lithosphere carries the largest quantity of all earth's oxygen, about 99.5%, because it is a constituent of the earth's lands, soils, organic matter, biomass, water, and rocks. Mostly, these constituents of the earth fix oxygen in mineral chemicals compounds such as oxides and silicates. The process is natural and happens automatically as the pure mineral elements absorb or react with the free oxygen. It happens similar to the manner in which iron picks up oxygen from the air, resulting in the formation of rust (iron oxides).

As such, during chemical reactions and some weathering processes, a portion of the trapped oxygen in the minerals is released into the atmosphere. Also, as animals and plants draw nutrient minerals from rocks, organic matter, or biomass, some of the trapped oxygen is freed in the process. Dissolved oxygen is also present in water system which is essential for the survival of aquatic life forms. As a result, these processes combined gives rise to oxygen cycle in the biosphere and lithosphere.

Processes that Use Oxygen

1. Respiration: When we breather, we use oxygen and release carbon dioxide. Similarly, animals and plants also use oxygen when they breathe.

2. Combustion: When you burn something let's say a paper, you need three things for combustion to take place i.e. oxygen, fuel and heat. So, when you burn a paper, it uses oxygen and releases carbon dioxide and may be some other gases.

3. Decomposition: Decomposition occurs when plants and animals die. When this happens, they decompose and such process uses oxygen and releases carbon dioxide.

4. Rusting: When things rust, they use oxygen. This is also called as oxidation.

Processes that Produce Oxygen

1. Plants: Plants produce oxygen during the process of photosynthesis. During the process of photosynthesis,

plants use carbon dioxide, sunlight, and water to create energy.

2. Sunlight: Some oxygen also gets produced when sunlight reacts with water vapor in the atmosphere.

Nitrogen Cycle

The nitrogen cycle refers to the cycle of nitrogen atoms through the living and non-living systems of Earth. Because nitrogen is a necessary ingredient for life as we know it, the nitrogen cycle is vital to sustaining life on Earth.

Nitrogen was originally formed in the hearts of stars through the process of nuclear fusion. When ancient stars exploded, they flung nitrogen-containing gases across the Universe. When the Earth was formed, nitrogen gas was a main ingredient in its atmosphere.

Today, the Earth's atmosphere is about 78% nitrogen, about 21% oxygen, and about 1% other gases. This is an ideal balance because too much oxygen can actually be toxic to cells, as well as being highly flammable. Nitrogen, on the other hand, is inert and harmless in its gaseous form.

Here we will discuss how nitrogen plays a vital role in the chemistry of life – and how it gets from the atmosphere, into living things, and back again.

Function of Nitrogen Cycle

Nitrogen is an essential ingredient for life as we know it. Its unique chemical bonding properties allow it to create structures such as DNA and RNA nucleotides, and the amino acids from which proteins are built. Without nitrogen, these molecules would not be able to exist.

It's thought that the first nucleotides and amino acids formed naturally under the volatile conditions of early Earth, where energy sources like lightning strikes could cause nitrogen and other atoms to react and form complex structures

This process might have naturally produced self-replicating organic chemicals – but in order to reproduce and evolve, life needed to figure out how to make these nitrogen compounds on demand.

Today, "nitrogen fixers" are organisms that can turn nitrogen gas from the atmosphere into nitrogen compounds that other organisms can use to produce nucleic acids, amino acids, and more. These nitrogen fixers are such a vital part of the ecosystem that agriculture cannot occur without them.

Ancient peoples learned that if they did not alternate growing nitrogen-consuming crops with nitrogen-fixing crops, their farms would become fallow and unable to support growth. Today, most artificial fertilizers contain life-giving nitrogen compounds as their main ingredient to make soil more fertile.

Processes

Nitrogen is present in the environment in a wide variety of chemical forms including organic nitrogen, ammonium (NH_4^+), nitrite(NO_2^-), nitrate (NO_3^-), nitrous oxide (N_2O), nitric oxide (NO) or inorganic nitrogen gas (N_2). Organic nitrogen may be in the form of a living organism, humusor in the intermediate products of organic matter decomposition. The processes of the nitrogen cycle transform nitrogen from one form to another. Many of those processes are carried out by microbes, either in their effort to harvest energy or to accumulate nitrogen in a form needed for their growth. For example, the nitrogenous wastes in animal urine are broken down by nitrifying bacteria in the soil to be used as new. The diagram {besides} alongside shows how these processes fit together to form the nitrogen cycle.

Nitrogen Fixation

The conversion of nitrogen gas (N_2) into nitrates and nitrites through atmospheric, industrial and biological processes is called nitrogen fixation. Atmospheric nitrogen must be processed, or "fixed", into a usable form to be taken up by plants. Between $5x10^{12}$ and $10x10^{12}$ g per year are fixed by lightning strikes, but most fixation is done by free-living or symbiotic bacteria known as diazotrophs. These bacteria have the nitrogenase enzyme that combines gaseous nitrogen with hydrogen to produce ammonia, which is converted by the bacteria into other organic compounds. Most biological nitrogen fixation occurs by the activity of Mo-nitrogenase, found in a wide variety of bacteria and some Archaea. Mo-nitrogenase is a complex two-component enzyme that has multiple metal-containing prosthetic groups. An example of free-living bacteria is Azotobacter. Symbiotic nitrogen-fixing bacteria such as Rhizobium usually live in the root nodules of legumes (such as peas, alfalfa, and locust trees). Here they form a mutualistic relationship with the plant, producing ammonia in exchange for carbohydrates. Because of this relationship, legumes will often increase the nitrogen content of nitrogen-poor soils. A few non-legumes can also form such symbioses. Today, about 30% of the total fixed nitrogen is produced industrially using the Haber-Bosch process, which uses high temperatures and pressures to convert nitrogen gas and a hydrogen source (natural gas or petroleum) into ammonia.

Assimilation

Plants can absorb nitrate or ammonium from the soil via their root hairs. If nitrate is absorbed, it is first reduced to nitrite ions and then ammonium ions for incorporation into amino acids, nucleic acids, and chlorophyll. In plants that have a symbiotic relationship with rhizobia, some nitrogen is assimilated in the form of ammonium ions directly from the nodules. It is now known that there is a more complex cycling of amino acids between Rhizobia bacteroids and plants. The plant provides amino acids to the bacteroids so ammonia assimilation is not required and the bacteroids pass amino acids (with the newly fixed nitrogen) back to the plant, thus forming an interdependent relationship. While many animals, fungi, and other heterotrophicorganisms obtain nitrogen by ingestion of amino acids, nucleotides, and other small organic molecules, other heterotrophs (including many bacteria) are able to utilize inorganic compounds, such as ammonium as sole N sources. Utilization of various N sources is carefully regulated in all organisms.

Ammonification

When a plant or animal dies or an animal expels waste, the initial form of nitrogen is organic. Bacteria or fungi convert the organic nitrogen within the remains back into ammonium (NH_4^+), a process called ammonification or mineralization. Enzymes involved are:

- GS: Gln Synthetase (Cytosolic & Plastic)
- GOGAT: Glu 2-oxoglutarate aminotransferase (Ferredoxin & NADH-dependent)
- GDH: Glu Dehydrogenase:

o Minor Role in ammonium assimilation.

o Important in amino acid catabolism.

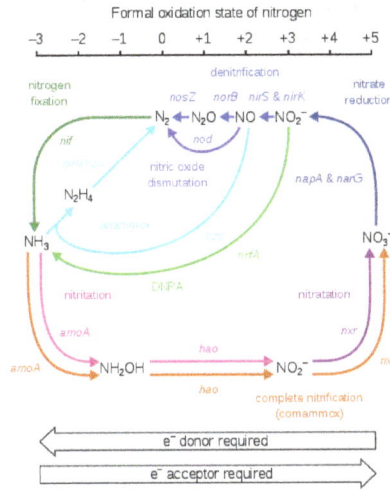

A schematic representation of the microbial nitrogen cycle. ANAMMOX is anaerobic ammonium oxidation, DNRA is dissimilatory nitrate reduction to ammonium, and COMMAMOXis complete ammonium oxidation.

The conversion of ammonium to nitrate is performed primarily by soil-living bacteria and other nitrifying bacteria. In the primary stage of nitrification, the oxidation of ammonium (NH_4^+) is performed by bacteria such as the Nitrosomonas species, which converts ammonia to nitrites (NO_2^-). Other bacterial species such as Nitrobacter, are responsible for the oxidation of the nitrites (NO_2^-) into nitrates (NO_3^-). It is important for the ammonia (NH_3) to be converted to nitrates or nitrites because ammonia gas is toxic to plants.

Due to their very high solubility and because soils are highly unable to retain anions, nitrates can enter groundwater. Elevated nitrate in groundwater is a concern for drinking water use because nitrate can interfere with blood-oxygen levels in infants and cause methemoglobinemia or blue-baby syndrome. Where groundwater recharges stream flow, nitrate-enriched groundwater can contribute to eutrophication, a process that leads to high algal population and growth, especially blue-green algal populations. While not directly toxic to fish life, like ammonia, nitrate can have indirect effects on fish if it contributes to this eutrophication. Nitrogen has contributed to severe eutrophication problems in some water bodies. Since 2006, the application of nitrogen fertilizer has been increasingly controlled in Britain and the United States. This is occurring along the same lines as control of phosphorus fertilizer, restriction of which is normally considered essential to the recovery of eutrophied waterbodies.

Denitrification

Denitrification is the reduction of nitrates back into nitrogen gas (N_2), completing the nitrogen cycle. This process is performed by bacterial species such as Pseudomonas and Clostridium in anaerobic conditions. They use the nitrate as an electron acceptor in the place of oxygen during respiration. These facultatively anaerobic bacteria can also live in aerobic conditions. Denitrification happens in anaerobic conditions e.g. waterlogged soils. The denitrifying bacteria use nitrates in the soil to carry out respiration and consequently produce nitrogen gas, which is inert and unavailable to plants.

Dissimilatory Nitrate Reduction to Ammonium

Dissimilatory nitrate reduction to ammonium (DNRA), or nitrate/nitrite ammonification, is an anaerobic respiration process. Microbes which undertake DNRA oxidise organic matter and use nitrate as an electron acceptor, reducing it to nitrite, then ammonium ($NO_3^- \rightarrow NO_2^- \rightarrow NH_4^+$). Both denitrifying and nitrate ammonification bacteria will be competing for nitrate in the environment, although DNRA acts to conserve bioavailable nitrogen as soluble ammonium rather than producing dinitrogen gas.

Anaerobic Ammonia Oxidation

In this biological process, nitrite and ammonia are converted directly into molecular nitrogen (N_2) gas. This process makes up a major proportion of nitrogen conversion in the oceans. The balanced formula for this "anammox" chemical reaction is: $NH_4^+ + NO_2^- \rightarrow N_2 + 2H_2O$ ($\Delta G^\circ = -357$ kJ mol^{-1}).

Other Processes

Though nitrogen fixation is the primary source of plant-available nitrogen in most ecosystems, in areas with nitrogen-rich bedrock, the breakdown of this rock also serves as a nitrogen source.

Marine Nitrogen Cycle

The nitrogen cycle is an important process in the ocean as well. While the overall cycle is similar, there are different players and modes of transfer for nitrogen in the ocean. Nitrogen enters the water through the precipitation, runoff, or as N_2 from the atmosphere. Nitrogen cannot be utilized by phytoplankton as N_2 so it must undergo nitrogen fixation which is performed predominately by cyanobacteria. Without supplies of fixed nitrogen entering the marine cycle, the fixed nitrogen would be used up in about 2000 years.Phytoplankton need nitrogen in biologically available forms for the initial synthesis of organic matter. Ammonia and urea are released into the water by excretion from plankton. Nitrogen sources are removed from the euphotic zone by the downward movement of the organic matter. This can occur from sinking of phytoplankton, vertical mixing, or sinking of waste of vertical migrators. The sinking results in ammonia being introduced at lower depths below the euphotic zone. Bacteria are able to convert ammonia to nitrite and nitrate but they are inhibited by light so this must occur below the euphotic zone. Ammonification or Mineralization is performed by bacteria to convert organic nitrogen to ammonia. Nitrification can then occur to convert the ammonium to nitrite and nitrate. Nitrate can be returned to the euphotic zone by vertical mixing and upwelling where it can be taken up by phytoplankton to continue the cycle. N_2 can be returned to the atmosphere through denitrification.

Ammonium is thought to be the preferred source of fixed nitrogen for phytoplankton because its assimilation does not involve a redox reaction and therefore requires little energy. Nitrate requires a redox reaction for assimilation but is more abundant so most phytoplankton have adapted to have the enzymes necessary to undertake this reduction (nitrate reductase). There are a few notable and well-known exceptions that include Prochlorococcus and some Synechococcus. These species can only take up nitrogen as ammonium.

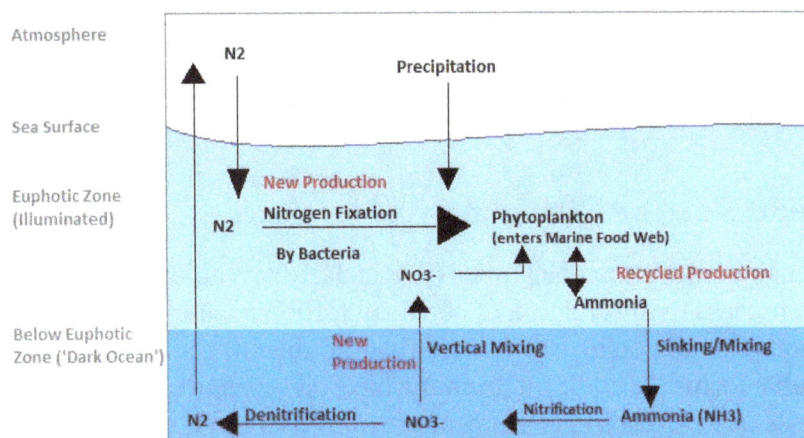

A schematic representing the Marine Nitrogen Cycle

The nutrients in the ocean are not uniformly distributed. Areas of upwelling provide supplies of nitrogen from below the euphotic zone. Coastal zones provide nitrogen from runoff and upwelling occurs readily along the coast. However, the rate at which nitrogen can be taken up by phytoplankton is decreased in oligotrophic waters year-round and temperate water in the summer resulting in lower primary production. The distribution of the different forms of nitrogen varies throughout the oceans as well.

Nitrate is depleted in near-surface water except in upwelling regions. Coastal upwelling regions usually have high nitrate and chlorophyll levels as a result of the increased production. However, there are regions of high surface nitrate but low chlorophyll that are referred to as HNLC (high nitrogen, low chlorophyll) regions. The best explanation for HNLC regions relates to iron scarcity in the ocean, which may play an important part in ocean dynamics and nutrient cycles. The input of iron varies by region and is delivered to the ocean by dust (from dust storms) and leached out of rocks. Iron is under consideration as the true limiting element to ecosystem productivity in the ocean.

Ammonium and nitrite show a maximum concentration at 50–80 m (lower end of the euphotic zone) with decreasing concentration below that depth. This distribution can be accounted for by the fact that nitrite and ammonium are intermediate species. They are both rapidly produced and consumed through the water column. The amount of ammonium in the ocean is about 3 orders of magnitude less than nitrate. Between ammonium, nitrite, and nitrate, nitrite has the fastest turnover rate. It can be produced during nitrate assimilation, nitrification, and denitrification; however, it is immediately consumed again.

New vs. Regenerated Nitrogen

Nitrogen entering the euphotic zone is referred to as new nitrogen because it is newly arrived from outside the productive layer. The new nitrogen can come from below the euphotic zone or from outside sources. Outside sources are upwelling from deep water and nitrogen fixation. If the organic matter is eaten, respired, delivered to the water as ammonia, and re-incorporated into organic matter by phytoplankton it is considered recycled/regenerated production. New production is an important component of the marine environment. One reason is that only continual input of new nitrogen can determine the total capacity of the ocean to produce a sustainable fish harvest. Harvesting fish from regenerated nitrogen areas will lead to a decrease in nitrogen and therefore a decrease in primary production. This will have a negative effect on the system. However, if fish are harvested from areas of new nitrogen the nitrogen will be replenished.

Human Influences on the Nitrogen Cycle

As a result of extensive cultivation of legumes (particularly soy, alfalfa, and clover), growing use of the Haber–Bosch process in the creation of chemical fertilizers, and pollution emitted by vehicles and industrial plants, human beings have more than doubled the annual transfer of nitrogen into biologically available forms. In addition, humans have significantly contributed to the transfer of nitrogen trace gases from Earth to the atmosphere and from the land to aquatic systems. Human alterations to the global nitrogen cycle are most intense in developed countries and in Asia, where vehicle emissions and industrial agriculture are highest.

Nitrous oxide (N_2O) has risen in the atmosphere as a result of agricultural fertilization, biomass burning, cattle and feedlots, and industrial sources. N_2O has deleterious effects in the stratosphere, where it breaks down and acts as a catalyst in the destruction of atmospheric ozone. Nitrous oxide is also a greenhouse gas and is currently the third largest contributor to global warming, after carbon dioxide and methane. While not as abundant in the atmosphere as carbon dioxide, it is, for an equivalent mass, nearly 300 times more potent in its ability to warm the planet.

Ammonia (NH_3) in the atmosphere has tripled as the result of human activities. It is a reactant in the atmosphere, where it acts as an aerosol, decreasing air quality and clinging to water droplets, eventually resulting in nitric acid (HNO_3) that produces acid rain. Atmospheric ammonia and nitric acid also damage respiratory systems.

The very high temperature of lightning naturally produces small amounts of NO_x, NH_3, and HNO_3, but high-temperature combustion has contributed to a 6- or 7-fold increase in the flux of NO_x to the atmosphere. Its production is a function of combustion temperature - the higher the temperature, the more NO_x is produced. Fossil fuel combustion is a primary contributor, but so are biofuels and even the burning of hydrogen. However, the rate that hydrogen is directly injected into the combustion chambers of internal combustion engines can be controlled to prevent the higher combustion temperatures that produce NO_x.

Ammonia and nitrous oxides actively alter atmospheric chemistry. They are precursors of tropospheric (lower atmosphere) ozone production, which contributes to smog and acid rain, damages plants and increases nitrogen inputs to ecosystems. Ecosystem processes can increase with nitrogen fertilization, but anthropogenic input can also result in nitrogen saturation, which weakens productivity and can damage the health of plants, animals, fish, and humans.

Decreases in biodiversity can also result if higher nitrogen availability increases nitrogen-demanding grasses, causing a degradation of nitrogen-poor, species-diverse heathlands.

Environmental Impacts

Increasing levels of nitrogen deposition are shown to have a number of negative effects on both terrestrial and aquatic ecosystems. Nitrogen gases and aerosols can be directly toxic to certain plant species, affecting the aboveground physiology and growth of plants near large point sources of nitrogen pollution. Changes to plant species interactions may also occur, as accumulation of nitrogen compounds increase its availability in a given ecosystem, eventually changing the species composition, plant diversity, and nitrogen cycling. Ammonia and ammonium - two reduced forms of nitrogen - can be detrimental over time due to an increased toxicity toward sensitive species of plants, particularly those that are accustomed to using nitrate as their source of nitrogen, causing poor development of their roots and shoots. Increased nitrogen deposition also leads to soil acidification, which increases base cation leaching in the soil and amounts of aluminium and other potentially toxic metals, along with decreasing the amount of nitrification occurring and increasing plant-derived litter. Due to the ongoing changes caused by high nitrogen deposition, an environment›s susceptibility to ecological stress and disturbance - such as pests and pathogens - may increase, thus making it less resilient to situations that otherwise would have little impact to its long-term vitality.

Additional risks posed by increased availability of inorganic nitrogen in aquatic ecosystems include water acidification; eutrophication of fresh and saltwater systems; and toxicity issues for animals, including humans. Eutrophication often leads to lower dissolved oxygen levels in the water column, including hypoxic and anoxic conditions, which can cause death of aquatic fauna. Relatively sessile benthos, or bottom-dwelling creatures, are particularly vulnerable because of their lack of mobility, though large fish kills are not uncommon. Oceanic dead zones near the mouth of the Mississippi in the Gulf of Mexico are a well-known example of algal bloom-induced hypoxia. The New York Adirondack Lakes, Catskills, Hudson Highlands, Rensselaer Plateau and parts of Long Island display the impact of nitric acid rain deposition, resulting in the killing of fish and many other aquatic species.

Ammonia (NH_3) is highly toxic to fish and the level of ammonia discharged from wastewater treatment facilities must be closely monitored. To prevent fish deaths, nitrification via aeration prior to discharge is often desirable. Land application can be an attractive alternative to the aeration.

What is the importance of the nitrogen cycle?

- As we all know by now, the nitrogen cycle helps bring in the inert nitrogen from the air into the biochemical process in plants and then to animals.

- Plants need nitrogen to synthesise chlorophyll and so the nitrogen cycle is absolutely essential for them.

- During the process of ammonification, the bacteria help degrade decomposing animal and plant matter. This helps in naturally cleaning up the environment.

- Due to the nitrogen cycle, nitrates and nitrites are released into the soil which helps in enriching the soil with nutrients needed for cultivation.

- As plants use nitrogen for their biochemical processes, animals obtain the nitrogen and nitrogen compounds from plants. Nitrogen is needed as is an integral part of the cell composition. It is due to the nitrogen cycle that animals are also able to utilise the nitrogen present in the air.

Phosphorus Cycle

Phosphorus is an important element for all forms of life. As phosphate (PO4), it makes up an important part of the structural framework that holds DNA and RNA together. Phosphates are also a critical component of ATP—the cellular energy carrier—as they serve as an energy ?release' for organisms to use in building proteins or contacting muscles. Like calcium, phosphorus is important to vertebrates; in the human body, 80% of phosphorous is found in teeth and bones.

The phosphorus cycle differs from the other major biogeochemical cycles in that it does not include a gas phase; although small amounts of phosphoric acid (H3PO4) may make their way into the atmosphere, contributing—in some cases—to acid rain. The water, carbon, nitrogen and sulfur cycles all include at least one phase in which the element is in its gaseous state. Very little phosphorus circulates in the atmosphere because at Earth's normal temperatures and pressures, phosphorus and its various compounds are not gases. The largest reservoir of phosphorus is in sedimentary rock.

It is in these rocks where the phosphorus cycle begins. When it rains, phosphates are removed from the rocks (via weathering) and are distributed throughout both soils and water. Plants take up the phosphate ions from the soil. The phosphates then moves from plants to animals when herbivores eat plants and carnivores eat plants or herbivores. The phosphates absorbed by animal tissue through consumption eventually returns to the soil through the excretion of urine and feces, as well as from the final decomposition of plants and animals after death.

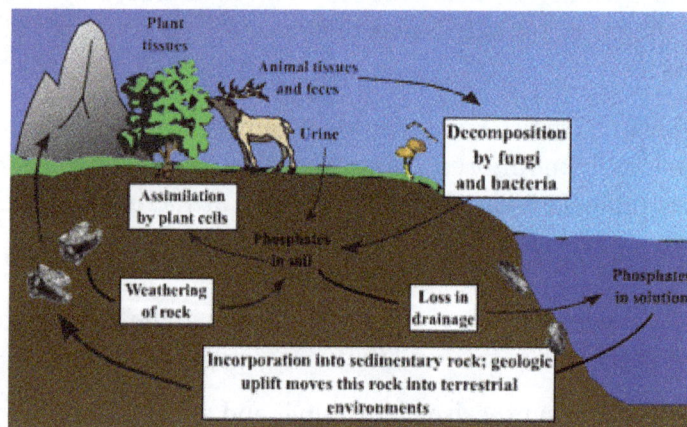

The same process occurs within the aquatic ecosystem. Phosphorus is not highly soluble, binding tightly to molecules in soil, therefore it mostly reaches waters by traveling with runoff soil particles. Phosphates also enter waterways through fertilizer runoff, sewage seepage, natural mineral deposits, and wastes from other industrial processes. These phosphates tend to settle on ocean floors and lake bottoms. As sediments are stirred up, phosphates may reenter the phosphorus cycle, but they are more commonly made available to aquatic organisms by being exposed through erosion. Water plants take up the waterborne phosphate which then travels up through successive stages of the aquatic food chain.

While obviously beneficial for many biological processes, in surface waters an excessive concentration of phosphorus is considered a pollutant. Phosphate stimulates the growth of plankton and plants, favoring weedy species over others. Excess growth of these plants tend to consume large amounts of dissolved oxygen, potentially suffocating fish and other marine animals, while also blocking available sunlight to bottom dwelling species. This is known as eutrophication.

Phosphorus in the Environment

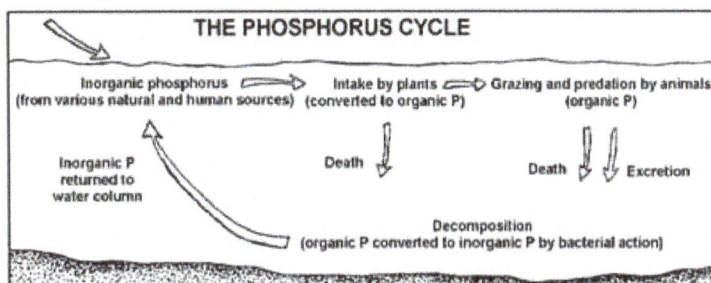

The aquatic phosphorus cycle

Ecological Function

Phosphorus is an essential nutrient for plants and animals. Phosphorus is a limiting nutrient for aquatic organisms. Phosphorus forms parts of important life-sustaining molecules that are very common in the biosphere. Phosphorus does not enter the atmosphere, remaining mostly on land and in rock and soil minerals. Eighty percent of the mined phosphorus is used to make fertilizers. Phosphates from fertilizers, sewage and detergents can cause pollution in lakes and streams. Overenrichment of phosphate in both fresh and inshore marine waters can lead to massive algae blooms which, when they die and decay, leads to eutrophication of fresh waters only. An example of this is the Canadian Experimental Lakes Area. These freshwater algal blooms should not be confused with those in saltwater environments. Recent research suggests that the predominant pollutant responsible for algal blooms in salt water estuaries and coastal marine habitats is Nitrogen.

Phosphorus occurs most abundantly in nature as part of the orthophosphate ion $(PO_4)^{3-}$, consisting of a P atom and 4 oxygen atoms. On land most phosphorus is found in rocks and minerals. Phosphorus rich deposits have generally formed in the ocean or from guano, and over time, geologic processes bring ocean sediments to land. Weathering of rocks and minerals release phosphorus in a soluble form where it is taken up by plants, and it is transformed into organic compounds. The plants may then be consumed by herbivores and the phosphorus is either incorporated into their tissues or excreted. After death, the animal or plant decays, and phosphorus is returned to the soil where a large part of the phosphorus is transformed into insoluble compounds. Runoff may carry a small part of the phosphorus back to the ocean. Generally with time (thousands of years) soils become deficient in phosphorus leading to ecosystem retrogression.

Biological Function

The primary biological importance of phosphates is as a component of nucleotides, which serve as energy storage within cells (ATP) or when linked together, form the nucleic acids DNA and RNA. The double helix of our DNA is only possible because of the phosphate ester bridge that binds the helix. Besides making biomolecules, phosphorus is also found in bone and the enamel of mammalian teeth, whose strength is derived from calcium phosphate in the form of Hydroxylapatite. It is also found in the exoskeleton of insects, and phospholipids (found in all biological membranes). It also functions as a buffering agent in maintaining acid base homeostasis in the human body.

Process of the Cycle

Phosphates move quickly through plants and animals; however, the processes that move them through the soil or ocean are very slow, making the phosphorus cycle overall one of the slowest biogeochemical cycles.

Initially, phosphate weathers from rocks and minerals, the most common mineral being apatite. Overall small losses occur in terrestrial environments by leaching and erosion, through the action of rain. In soil, phosphate is adsorbed on iron oxides, aluminium hydroxides, clay surfaces, and organic matter particles, and becomes incorporated (immobilized or fixed). Plants and fungi can also be active in making P soluble.

Unlike other cycles, P is not normally found in the air as a gas; it only occurs under highly reducing conditions as the gas phosphine PH_3.

Phosphatic Minerals

The availability of phosphorus in an ecosystem is restricted by the rate of release of this element during weathering. The release of phosphorus from apatite dissolution is a key control on ecosystem productivity. The primary mineral with significant phosphorus content, apatite $[Ca_5(PO_4)_3OH]$ undergoes carbonation.

Little of this released phosphorus is taken up by biota (organic form), whereas a larger proportion reacts with other soil minerals. This leads to precipitation into unavailable forms in the later stage of weathering and soil development. Available phosphorus is found in a biogeochemical cycle in the upper soil profile, while phosphorus found at lower depths is primarily involved in geochemical reactions with secondary minerals. Plant growth depends on the rapid root uptake of phosphorus released from dead organic matter in the biochemical cycle. Phosphorus is limited in supply for plant growth. Phosphates move quickly through plants and animals; however, the processes that move them through the soil or ocean are very slow, making the phosphorus cycle overall one of the slowest biogeochemical cycles.

Low-molecular-weight (LMW) organic acids are found in soils. They originate from the activities of various microorganisms in soils or may be exuded from the roots of living plants. Several of those organic acids are capable of forming stable organo-metal complexes with various metal ions found in soil solutions. As a result, these processes may lead to the release of inorganic phosphorus associated with aluminium, iron, and calcium in soil minerals. The production and release of oxalic acid by mycorrhizal fungi explain their importance in maintaining and supplying phosphorus to plants.

The availability of organic phosphorus to support microbial, plant and animal growth depends on the rate of their degradation to generate free phosphate. There are various enzymes such as phosphatases, nucleases and phytase involved for the degradation. Some of the abiotic pathways in the environment studied are hydrolytic reactions and photolytic reactions. Enzymatic hydrolysis of organic phosphorus is an essential step in the biogeochemical phosphorus cycle, including the phosphorus nutrition of plants and microorganisms and the transfer of organic phosphorus from soil to bodies of water. Many organisms rely on the soil derived phosphorus for their phosphorus nutrition.

Human Influences

Nutrients are important to the growth and survival of living organisms, and hence, are essential for development and maintenance of healthy ecosystems. Humans have greatly influenced the phosphorus cycle by mining phosphorus, converting it to fertilizer, and by shipping fertilizer and products around the globe. Transporting phosphorus in food from farms to cities has made a major change in the global Phosphorus cycle. However, excessive amounts of nutrients, particularly phosphorus and nitrogen, are detrimental to aquatic ecosystems. Waters are enriched in phosphorus from farms' run-off, and from effluent that is inadequately treated before it is discharged to waters. Natural eutrophication is a process by which lakes gradually age and become more productive and may take thousands of years to progress. Cultural or anthropogenic eutrophication, however, is water pollution caused by excessive plant nutrients; this results in excessive growth in the algal population; when this algae dies its putrefaction depletes the water of oxygen. Such eutrophication may also give rise to toxic algal bloom. Both these effects cause animal and plant death rates to increase as the plants take in poisonous water while the animals drink the poisoned water. Surface and subsurface runoff and erosion from high-phosphorus soils may be major contributing factors to this fresh water eutrophication. The processes controlling soil Phosphorus release to surface runoff and to subsurface flow are a complex interaction between the type of phosphorus input, soil type and management, and transport processes depending on hydrological conditions.

Repeated application of liquid hog manure in excess to crop needs can have detrimental effects on soil phosphorus status. Also, application of biosolids may increase available phosphorus in soil. In poorly drained soils or in areas where snowmelt can cause periodic waterlogging, dereducing conditions can be attained in 7–10 days. This causes a sharp increase in phosphorus concentration in solution and phosphorus can be leached. In addition, reduction of the soil causes a shift in phosphorus from resilient to more labile forms. This could eventually increase the potential for phosphorus loss. This is of particular concern for the environmentally sound management of such areas, where disposal of agricultural wastes has already become a problem. It is suggested that the water regime of soils that are to be used for organic wastes disposal is taken into account in the preparation of waste management regulations.

Human interference in the phosphorus cycle occurs by overuse or careless use of phosphorus fertilizers. This results in increased amounts of phosphorus as pollutants in bodies of water resulting in eutrophication. Eutrophication devastates water ecosystems by inducing anoxic conditions.

Sulfur Cycle

Sulfur (S), the tenth most abundant element in the universe, is a brittle, yellow, tasteless, and odorless non-metallic element. It comprises many vitamins, proteins, and hormones that play critical roles in both climate and in the health of various ecosystems. The majority of the Earth's sulfur is stored underground in rocks and minerals, including as sulfate salts buried deep within ocean sediments.

The sulfur cycle contains both atmospheric and terrestrial processes. Within the terrestrial portion, the cycle begins with the weathering of rocks, releasing the stored sulfur. The sulfur then comes into contact with air where it is converted into sulfate (SO_4). The sulfate is taken up by plants and microorganisms and is converted into organic forms; animals then consume these organic forms through foods they eat, thereby moving the sulfur through the food chain. As organisms die and decompose, some of the sulfur is again released as a sulfate and some enters the tissues of microorganisms. There are also a variety of natural sources that emit sulfur directly into the atmosphere, including volcanic eruptions, the breakdown of organic matter in swamps and tidal flats, and the evaporation of water.

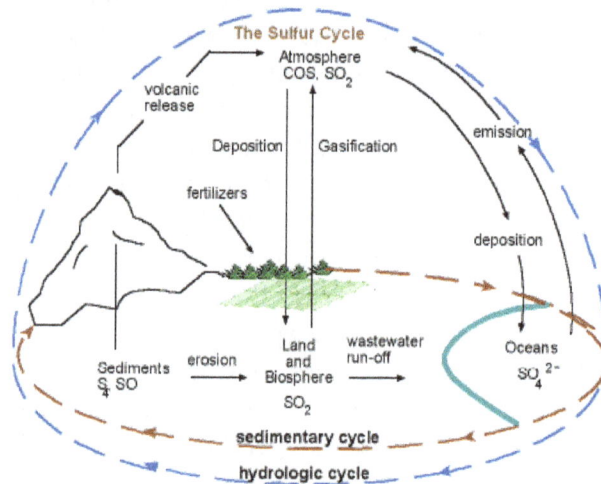

Sulfur eventually settles back into the Earth or comes down within rainfall. A continuous loss of sulfur from terrestrial ecosystem runoff occurs through drainage into lakes and streams, and eventually oceans. Sulfur also enters the ocean through fallout from the Earth's atmosphere. Within the ocean, some sulfur cycles through marine communities, moving through the food chain. A portion of this sulfur is emitted back into the atmosphere fromd sea spray. The remaining sulfur is lost to the ocean depths, combining with iron to form ferrous sulfide which is responsible for the black color of most marine sediments.

Since the Industrial Revolution, human activities have contributed to the amount of sulfur that enters the atmosphere, primarily through the burning of fossil fuels and the processing of metals. One-third of all sulfur that reaches the atmosphere—including 90% of sulfur dioxide—stems from human activities. Emissions from these activities, along with nitrogen emissions, react with other chemicals in the atmosphere to produce tiny particles of sulfate salts which fall as acid rain, causing a variety of damage to both the natural environment as well as to man-made environments, such as the chemical weathering of buildings. However, as particles and tiny airborne droplets, sulfur also acts as a regulator of global climate. Sulfur dioxide and sulfate aerosols absorb ultraviolet radiation, creating cloud cover that cools cities and may offset global warming caused by the greenhouse effect. The actual amount of this offset is a question that researchers are attempting to answer.

Most of the earth's sulphur is tied up in rocks and salts or buried deep in the ocean in oceanic sediments. Sulphur can also be found in the atmosphere. It enters the atmosphere through both natural and human sources. Natural recourses can be for instance volcanic eruptions, bacterial processes, evaporation from water, or decaying organisms. When sulphur enters the atmosphere through human activity, this is mainly a consequence of industrial processes where sulphur dioxide (SO_2) and hydrogen sulphide (H_2S) gases are emitted on a wide scale. When sulphur dioxide enters the atmosphere it will react with oxygen to produce sulphur trioxide gas (SO_3), or with other chemicals in the atmosphere, to produce sulphur salts. Sulphur dioxide may also react with water to produce sulphuric acid (H_2SO_4). Sulphuric acid may also be produced from demethylsulphide, which is emitted to the atmosphere by plankton species. All these particles will settle back onto earth, or react with rain and fall back onto earth as acid deposition. The particles will than be absorbed by plants again and are released back into the atmosphere, so that the sulphur cycle will start over again.

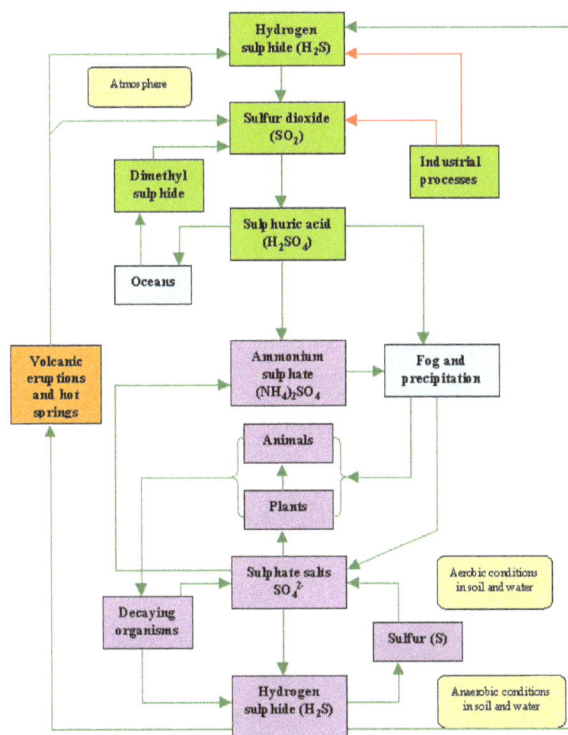

A schematic representation of the sulphur cycle

References

- Baedke, Steve J.; Fichter, Lynn S. "Biogeochemical Cycles: Carbon Cycle". Supplimental Lecture Notes for Geol 398. James Madison University. Retrieved 20 November 2017

- Rudolf Dvořák (2007). Extrasolar Planets. Wiley-VCH. pp. 139–140. ISBN 978-3-527-40671-5. Retrieved 2009-05-05

- Britto, Dev T.; Kronzucker, Herbert J. "NH4+ toxicity in higher plants: a critical review". Journal of Plant Physiology. 159 (6): 567–584. doi:10.1078/0176-1617-0774

- "Mercury Cycling in the Environment". Wisconsin Water Science Center. United States Geological Survey. 10 January 2013. Retrieved 20 November 2017

- Metzger, Bruce M.; Coogan, Michael D. (1993). The Oxford Companion to the Bible. New York, NY: Oxford University Press. p. 369. ISBN 0195046455

- Marchant, H. K., Lavik, G., Holtappels, M., and Kuypers, M. M. M. (2014). "The Fate of Nitrate in Intertidal Permeable Sediments". PLoS ONE. 9 (8). doi:10.1371/journal.pone.0104517

- McGuire, 1A. D.; Lukina, N. V. (2007). "Biogeochemical cycles". In Groisman, P.; Bartalev, S. A.; NEESPI Science Plan Development Team. Northern Eurasia earth science partnership initiative (NEESPI), Science plan overview (PDF). Global Planetary Change. 56. pp. 215–234. Retrieved 20 November 2017

- Merrill, Eugene H.; Rooker, Mark F.; Grisanti, Michael A. (2011). The World and the Word. Nashville, TN: B&H Academic. p. 430. ISBN 9780805440317

- Paul J. Durack; Susan E. Wijffels & Richard J. Matear (27 April 2012). "Ocean Salinities Reveal Strong Global Water Cycle Intensification During 1950 to 2000". Science. 336 (6080): 455–458. Bibcode:2012Sci...336..455D. doi:10.1126/science.1212222. PMID 22539717

- Lal, Rattan (2008). "Sequestration of atmospheric CO_2 in global carbon pools". Energy and Environmental Science. 1: 86–100. doi:10.1039/b809492f

Permissions

Index

A

Abiotic Component, 42
Abiotic Factors, 15, 42-43, 173-174
Algal Blooms, 82, 203
Animal Populations, 1, 35, 39, 41
Applied Ecology, 36-38
Aquatic Ecosystem, 58-59, 62, 184, 202

B

Biogeochemical Cycles, 16, 19, 23-24, 102, 172-173, 175, 202, 204, 207-208
Biomass of Organisms, 95, 97, 125
Biome, 10, 21, 48, 108
Biotic Component, 44

C

Carnivores, 30, 45, 50, 70, 89-91, 94, 112-115, 117, 121, 123, 125, 202
Climate Change, 14, 27, 55, 58, 60, 80, 83, 85-86, 181, 187-189
Community Ecology, 7, 30
Conservation Biology, 1, 5, 21, 33-34, 37-38

D

Detritus Food Chain, 91-92

E

Earth's Atmosphere, 1, 13, 15-16, 24, 68, 185, 188, 195, 206
Ecological Niche, 3-4
Ecological Pyramid, 89, 93-94, 127-129
Ecological Theory, 10, 13, 27, 29, 53
Ecological Trap, 1, 46
Ecosystem Ecology, 7, 50, 89
Ecosystem Engineering, 4-5
Energy Flow, 49, 89-92, 96, 98, 115-118, 120, 125
Environmental Impacts, 55, 200
Eutrophication, 61, 81-82, 84, 182, 197, 201, 203, 205

F

Food Webs, 7-8, 50, 61, 112-117, 119-123, 125, 133, 190
Fragmentation, 35-36, 51, 58
Freshwater Ecosystems, 58-60

G

Geologic Time, 182, 190
Grazing Food Chain, 91-92, 113, 125
Groundwater, 75, 176-180, 182, 197

H

Habitats, 3-4, 7-8, 18, 23-24, 38-39, 41, 54, 59, 62, 64-65, 69, 72, 75, 78, 81, 87, 101, 110, 147, 165, 203
Herbivores, 7-9, 25, 31, 45, 50, 56, 63, 89-91, 94-97, 99, 112-117, 123-127, 202-203
Heterotrophic Organisms, 47, 63, 97
Hnlc, 108, 199
Homeostasis, 4, 163, 204
Human Ecological Theory, 27, 29
Human Influences, 79, 200, 205
Hydrologic Cycle, 175, 179, 181

K

K-selected Psychology, 168-169
K-selected Reproductive Strategy, 167, 169
Keystone Species, 8-10, 54, 80, 100, 118, 123

L

Lake Ecosystems, 60
Land Surface, 58, 177-178
Landscape Ecology, 31-36
Lentic, 60-61

M

Marine Ecosystems, 59-60, 63, 80-81, 183
Marine Life, 43, 64, 81
Metapopulation, 6, 35, 135, 143, 171
Microorganisms, 9, 17, 23, 28, 30, 45, 47, 51, 71, 74, 86, 101, 108, 115, 118-119, 204-206

N

Nitrogen Fixation, 52, 108, 196, 198-199

O

Omnivores, 8, 45, 56, 113, 115-116
Overfishing, 55, 60, 79-81
Oxygen Cycle, 175, 191-194
Ozone, 23, 65, 191, 193, 200

P

Parasitic Food Chain, 91-92, 94

Phosphorus Cycle, 104, 175, 202-205

Photosynthesis, 7, 10, 16-17, 19, 23, 26, 43-44, 47-49, 51-52, 65-66, 69, 75, 89, 96, 98-100, 102-106, 110, 112, 114, 127, 173, 183-185, 187, 189, 191-194

Plankton, 61, 68-73, 106-108, 198, 202, 207

Population Ecology, 5, 134-136, 143

Primary Production, 7, 9, 16, 49-52, 59-60, 73, 79, 96-99, 102, 104-111, 116, 119, 125, 132-133, 199-200

Production Rate, 97-98

R

R/k Selection Theory, 14, 134, 143, 163, 166, 168, 171

Reservoirs, 61, 84, 173-174, 178, 184-185, 191-192

S

Spatial Heterogeneity, 32-34

Substrate, 62, 71-72, 75-77, 105, 173

Sulfur Cycle, 174-175, 205-206

T

Terrestrial Ecosystems, 50, 52, 55-56, 80, 100, 105, 108, 119, 124-126, 183, 185, 188

Trophic Dynamics, 8-9, 116

U

Urban Ecology, 1, 39

Urban Ecosystem, 40

Urbanization, 160, 182

W

Waste Into the Sea, 80-81

Water Chemistry, 22, 76

Water Cycle, 22, 80, 173, 175-177, 179-183, 193-194, 208

Water Flow, 62, 74-75

Wetlands, 6, 19, 31-32, 40, 52, 60-63, 78, 86, 109, 115

www.ingramcontent.com/pod-product-compliance
Lightning Source LLC
Chambersburg PA
CBHW082033190326
41458CB00010B/3348